ISBN 978-1-330-47286-6
PIBN 10066365

AN ELEMENTARY TREATMENT OF THE THEORY OF SPINNING TOPS AND GYROSCOPIC MOTION

AN ELEMENTARY TREATMENT

OF THE THEORY OF

SPINNING TOPS

AND

GYROSCOPIC MOTION

BY

HAROLD CRABTREE, M.A.

FORMERLY SCHOLAR OF PEMBROKE COLLEGE, CAMBRIDGE
ASSISTANT MASTER AT CHARTERHOUSE

WITH ILLUSTRATIONS

LONGMANS, GREEN, AND CO.
39 PATERNOSTER ROW, LONDON
NEW YORK, BOMBAY, AND CALCUTTA
1909

GLASGOW: PRINTED AT THE UNIVERSITY PRESS
BY ROBERT MACLEHOSE AND CO. LTD.

TO MY OLD COACH AND FRIEND

DR. W. H. BESANT, F.R.S.

(*Senior Wrangler*, 1850)

The last of all your pupils I, the last
Of a long line, and some far-known to-day.
For me no fame: enough, the humbler way,
To follow you who pioneered the past.
And how we loved your book! wherein you cast
Your spells on moving worlds or meaner clay:—
The "jet of sand," the "falling chains," the sway
And poise of planet-spheres, dim, distant, vast.

You tell me now the wheels of Life run slow,
And the chains loosen. Sinks the lessening sand
While, as you rest, comes the calm evening glow,
"Cometh the night"—your motto. So you stand
To watch the day depart; yet, Master, know
Still burns your torch, passed on from hand to hand.

H. C.

1909.

PREFACE.

THE object of this book is to bring within the range of the abler Mathematicians at our Public Schools and of First Year undergraduates at the Universities, a subject which has hitherto been considered too difficult for any but the more advanced students in Mathematics, while even they have in many cases failed to derive more pleasure from the study of spinning tops than is contained in submitting the problem to the action of a complicated piece of Mathematical machinery which automatically, though unintelligently, turns out the correct result.

In this attempt to present an elementary, and at the same time a scientific view of the subject, I have expanded several of the suggestive ideas put forward in *Dynamics of Rotation* by Professor Worthington, to whom is due a large debt of thanks for much liberal criticism and assistance in the earlier chapters, and for permission to borrow his article on Brennan's Monorail.

In my other critics I have been similarly fortunate. Dr. Besant, Fellow of St. John's College, Cambridge, read and commented on several chapters in their first edition of manuscript; and in dedicating this volume to him I have endeavoured to express the interest in the subject, and the love of it, which I owe to his inspiring teaching. Professor Hopkinson of the Engineering Laboratory at Cambridge has also read certain portions, while much assistance has been rendered by my friend and late colleague Mr. W. M. Page, Fellow of King's College, Cambridge, with whom I have spent many pleasant hours discussing various points. Lastly, the whole work has been carefully criticised by Mr. G. Bistwistle, Fellow and Principal Mathematical Lecturer of Pembroke College, Cambridge, to whom my sincerest thanks are due for much time devoted to the book and for many valuable suggestions.

Considerable stress has been laid on Dimensions, particularly in the earlier pages, where quantities are equated which to the beginner appear at first sight to bear no relation to one another; and a series of "Questions" has been interspersed throughout, the answers to which, though requiring no calculation, ensure an intelligent grasp of the principles involved, and may induce the reader to think for himself of further side issues. It has been found convenient to distinguish the axis of symmetry of a top by the term *axle*, this being in general the only axis which is not passing through the material of the top, and whose motion involves a corresponding motion on the part of the top.

Particular care has been taken to preserve the elementary character of the first four chapters, which should be intelligible to any reader conversant with the principles of momentum and energy as applied to Particle Dynamics, while at the same time they embrace all that is necessary for understanding the motion of a spinning top, and the practical applications to the steering of a torpedo, the steadying of ships, and the monorail, which are given in Chapter V. Equations of motion, both steady and general, and a graphical representation of the paths described in space by the head of the top have been obtained directly from the principles of momentum, energy, and gyroscopic resistance. The latter has also been employed to establish the general equations of motion referred to moving axes, which are discussed subsequently and have been applied to the advanced part of the subject, including the oscillations of the sleeping top.

For great assistance in reading the proof sheets and for many valuable suggestions while going to Press, I have to thank my colleague Mr. W. A. Nayler; and also Mr. W. H. C. Romanis, now Minor Scholar of Trinity College, Cambridge, with whom I have conducted many experiments at Charterhouse, and to whom several of the earlier examples are due.

I also wish to acknowledge the courtesy and kindness of Messrs. Longmans, Green & Co., during the preparation of this volume.

Extreme care has been taken both in the arrangement and expression of ideas, but doubtless deficiencies yet remain, and any corrections or criticisms will be gratefully acknowledged.

HAROLD CRABTREE.

CHARTERHOUSE,
Feb. 1909.

CONTENTS.

INTRODUCTORY CHAPTER.

CHAPTER I.

ROTATION ABOUT A FIXED AXIS.

CHAPTER II.

REPRESENTATION OF ANGULAR VELOCITY. PRECESSION.

CHAPTER III.

DISCUSSION OF THE PHENOMENA DESCRIBED IN THE INTRODUCTORY CHAPTER.

CHAPTER IV.

THE STARTING OF PRECESSION. OSCILLATIONS OF THE GYROSCOPE.

CHAPTER V.

PRACTICAL APPLICATIONS.

CHAPTER VI.

STEADY MOTION OF A TOP.

CHAPTER VII.

GENERAL MOTION OF A TOP.

CHAPTER VIII.

MOVING AXES.

CHAPTER IX.

STABILITY OF ROTATION. PERIODS OF OSCILLATION.

INTRODUCTORY CHAPTER.

THERE are few of us who as boys have not been interested in Spinning Tops; but the day soon arrives when we become too old or too proud to spin them, and most of us from that time never give them a thought again. A few years later, perhaps, we become interested in some of the numerous inventions of modern days, or even look forward to exploring advanced regions of Mathematics and Science; but our tops have long since been forgotten. And yet, what surprising possibilities of knowledge and power have been put aside with our neglected playthings. Who would have thought that in them lay concealed the secret of steering a torpedo, of steadying ships, or of travelling with security in a single car on a tight-rope at the rate of 130 miles an hour? Last, but not least, who would have thought that on such motion as theirs depends the foundation of the most astounding theory of Modern Science that the so-called solid matter, with which we come in contact every day, is not really solid after all, but composed of a vast infinitude of particles whirling round one another at inconceivable speeds like planets in the Heavens.

Professor Perry, in his most fascinating book entitled *Spinning Tops*,* reminds us that an ordinary flexible metal chain or india-rubber band, which if left to itself would fall helplessly on the floor, will, if given a high speed of rotation on a revolving drum and then slipped off the drum, bounce along like an ordinary solid hoop; that a circular disc of paper if made to revolve very fast about an axis perpendicular to its plane will behave as if solid, and if struck with a stick make a loud resounding noise;† while, if rings of smoke are blown in a certain manner and made to collide, they will bounce off each other as if made of india-rubber. These instances begin to give

---

*The Romance of Science Series. S.P.C.K., price 2s. 6d.

† A comparatively small rate of rotation will enable a circular piece of ordinary writing paper to act as a circular saw, sufficiently powerful to cut another piece of paper right through.

It is not difficult to see the reason for this quasi-rigidity in the case of the flexible chain or the revolving paper disc. If we consider an element AB of the

us some faint idea of how what appears to be solid may in reality be made up of minute loose particles revolving round each other at an enormous rate: and it is interesting to think that nearly 2,000 years before this theory was formally stated, the Roman poet Lucretius should have written six books of majestic verse, one entirely, and the others partly, in support of a theory extremely similar to this, propounded by the Greek philosopher Democritus 400 years previously.

Let us consider for a few minutes the behaviour of an ordinary spinning top. It is full of surprising contradictions. In the first place, to take the most obvious of all, we cannot balance it on its peg; but give it a spin and it will stay balanced for a long time. We have known this all our lives; but few people can explain the reason, except to say "because it's spinning"—which begs the question. Again, supposing we spin it with its axis vertical and then give it a knock, it will go round the table in a slanting position; but if we spin it slanting to begin with, it will almost immediately stand upright. And once again, although it is the friction of the table and the resistance of the air which eventually bring a spinning top to rest, yet if we take a very smooth surface, such as a glass plate, we find that many tops will not spin on it at all.

The childish delight which we felt in watching our tops spinning remains with many of us a vivid recollection to this day. Some tops would buzz about busily before settling down to a regular motion; others would be steady and stately from the first. Some would "go to sleep" almost at once, and, if disturbed, would only show signs of life for a short time before going to sleep again: these when they "died" would die very suddenly. Others when disturbed would spin about in a slanting position for a long time, particularly those with rather a long "leg"; these took a long time to "die."

It is within everybody's experience that if a top is spun so that the foot traces out (approximately) a circle on the table, this circle will be described in the direction of rotation of the

---

chain we see that it is acted on (in addition to its weight) by two tensions, each of very great magnitude, since they act at a very small angle to each other and yet supply the normal force necessary for the circular motion. Hence any force which deflects AB from its circular position will have to be extremely large to

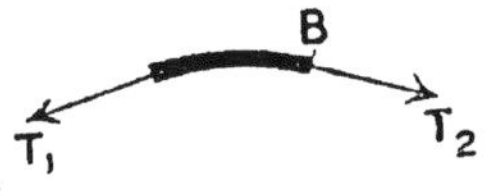

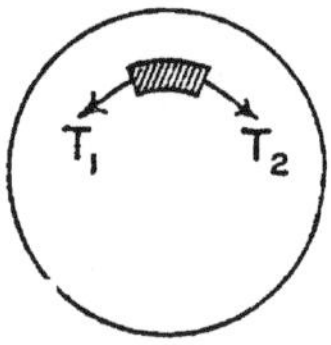

overcome the resolutes of these tensions which would be immediately called into play. Similarly in the case of the disc, any small elemental area is under the action of two very large tensions, resolutes of which are immediately called into play when a lateral blow is given to the disc.

top; in fact the whole top *rolls* round the circle leaning *towards* the centre of the circle; but the top in the accompanying figures

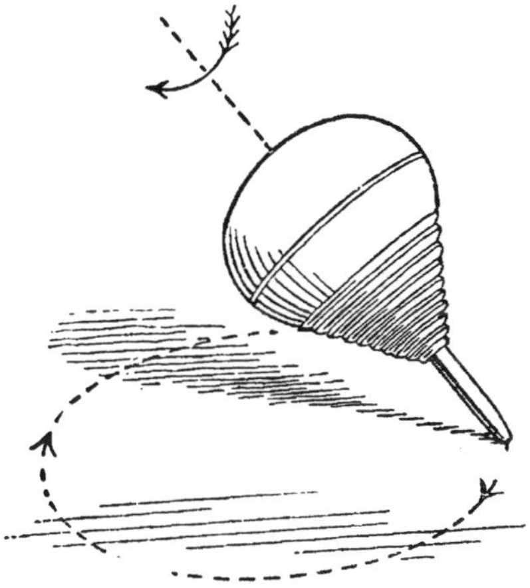

Fig. I.

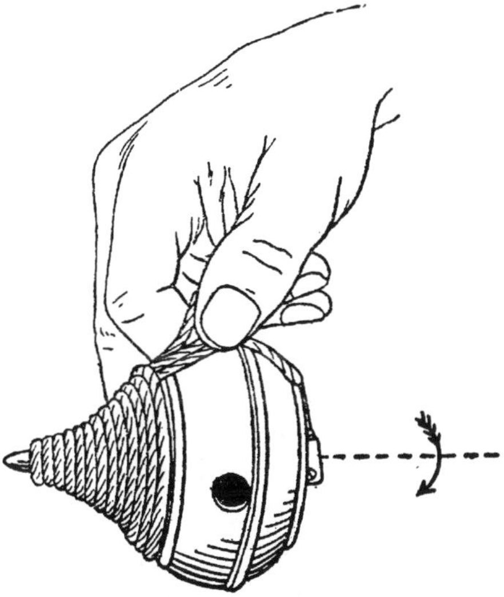

Fig. II.

will behave in exactly the opposite way. Here the string is fixed to the head by a swivel, and to spin the top the string

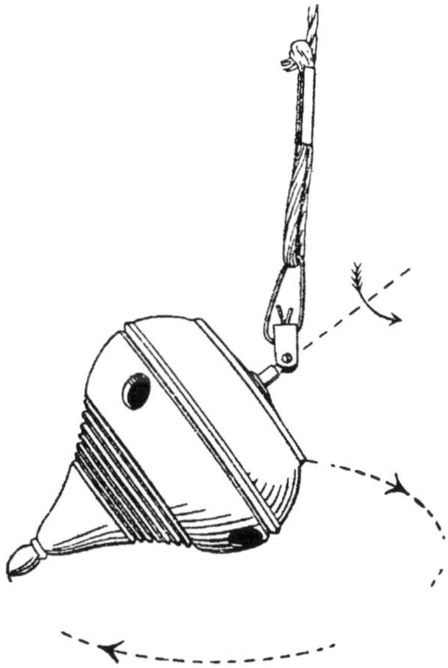

Fig. III. (*a*).

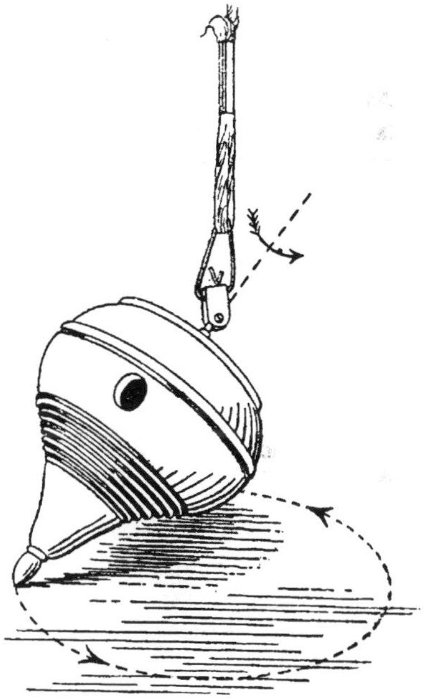

Fig. III. (*b*).

is pulled tight down to the toe and wound round, starting from the toe as for spinning an ordinary peg-top (Fig. II.). The top is then thrown down, when of course, instead of falling to the ground, it remains suspended from its head and spins, but instead of its toe (and its axis) moving round in the direction of the top's spin viewed, say, from above, it goes in the contrary direction (Fig. III. *a*). If the string is lowered so that the peg can touch the table (or floor) the toe of the top will immediately reverse its direction of motion (Fig. III. *b*); nor will the slightest jerk or effort be visible. The top will probably begin to rise to a more vertical position. If this motion is allowed to continue till the top begins to "die" and the axis becomes rather more horizontal, when it is lifted off the table by the string the immediate reversal of the direction of motion of the toe will become very obvious. The effect is a very pretty one, and it is interesting to note in passing that the Earth is a top of this description, *i.e.* one whose axis revolves (with a motion called "precession") in the opposite direction to the spin of rotation.

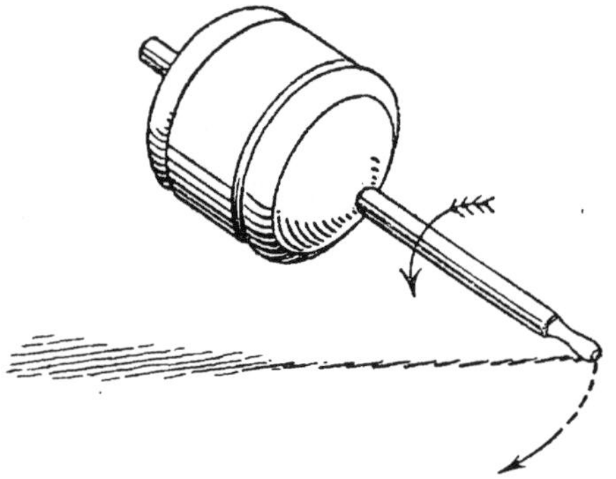

FIG. IV. (*a*)—Blunt peg.

The particular top represented in Figs. IV. (*a*) and IV. (*b*) is an ordinary whip-top through which has been inserted a metal spindle, which is capable of sliding movement through the top; so that either a very long or a very short leg can be used and the main body of the top can be either high up or low down. At the "foot" either a blunt or a fine peg can be screwed in, and it is found that, whether a long or short leg is used, the top will always rise more quickly from the slant position with the *blunt* peg. Any top will also rise more quickly on a *rough* surface than on a smooth one. If care is taken at the start to spin this top in a direction which should tighten the screw at the foot, it will be found when the top is "dead" that the screw is quite loose; but if the reverse spin be originally given this screw will end up tight. This top, if on its *fine* peg, maintains a steady motion for a long time when it is so nearly horizontal that the body nearly scrapes the table.

FIG. IV. (*b*).—Fine peg.

When the blunt peg is used this (practically) horizontal motion does not continue so long: and, in general, a top with a fine point or with a long leg will spin at a greater angle to the vertical than one with either a blunt point or short leg.

A loaded sphere when spun on a rough surface also presents a curious contradiction.

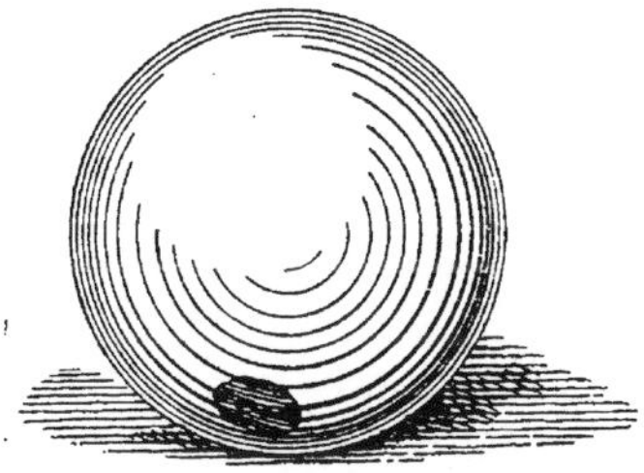

FIG. V. (*a*).

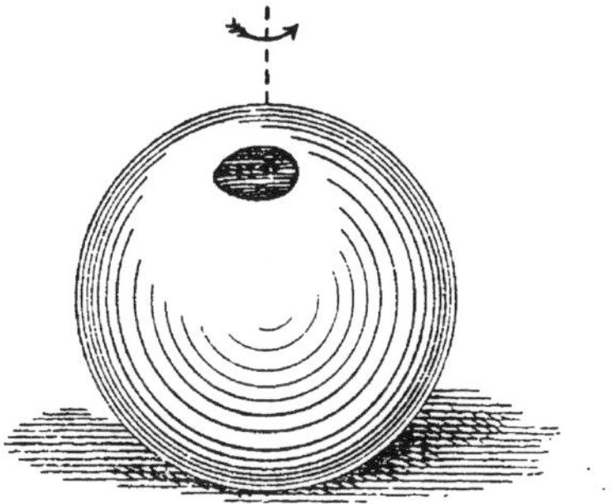

FIG. V. (*b*).

If, for example, a small hole is made in the side of a croquet ball and filled up with lead, when placed on a table the ball will settle down to the position where the lead touches the table (Fig. V. *a*). But if a really good spin be given to the ball the loaded part will persistently rise, as indicated in Fig. V. (*b*), and if the table is very rough it will get to the position where it is at the highest point of the sphere.

Many of the tops we have been accustomed to spin are hollow. Has it ever occurred to the reader to ask what would be the effect of filling a tin top with water, and making it watertight? The answer that occurs to most minds at once is probably that it would spin much better; it is heavier, and

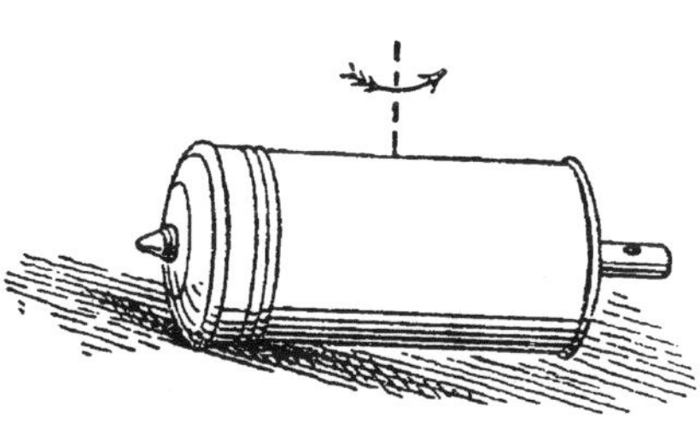

FIG. VI.—Top full.

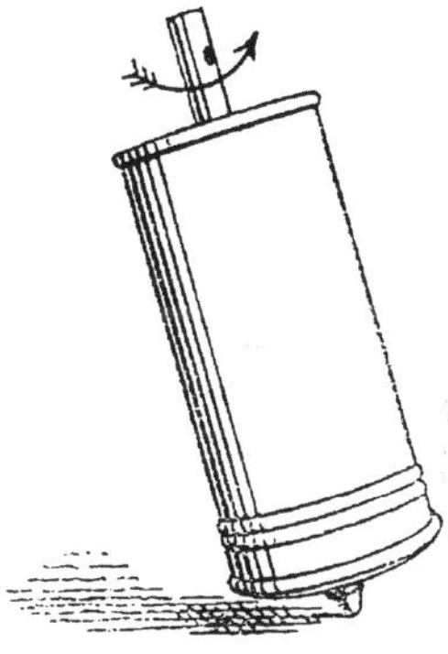

FIG. VII.—Top empty.

there is more of it generally. Let the experiment be made. Figs. VI., VII. represent two tops which are in every particular the same, except that the left hand one is full of water hermetically sealed. The empty one, if spun in the ordinary way, will continue to spin in an upright position; the other one will lie down on its side at once, and spin violently lying at full length on the table. Some such tops are a little uncertain which to do. That in Fig. VIII. has been constructed so that the head can be unscrewed and water poured in. If empty it spins very well, whether a big or little spin be originally given to it. When it is full of water a little spin will only result in the top falling to the ground; a good spin will keep it upright in spite of the water. Such tops can be readily constructed out of small tins or similar receptacles capable of being soldered and made water-tight.

FIG. VIII.

Figs. IX. and X. represent two hollow china eggs, exactly similar, with the exception that one has been filled with water and the hole stopped up with sealing wax. If they are laid down on the table, and a spin is given to each of them about a vertical axis, they will behave in entirely different ways. The empty one jumps up briskly on its end and continues to spin in that position for a long time; the other will spin slowly on its side for a short time and continue this uninteresting motion till it stops: or if stopped prematurely by laying a finger on it, will begin to spin again on removing the finger.

FIG. IX.—Egg full.

FIG. X.—Egg empty.

A real egg, if unboiled, and especially one which has had its yolk thoroughly shaken up into liquid form, will behave in precisely the same way; but a hard boiled egg will spin up on its end at once (especially if the table is rough) and continue spinning for a long time. A similar phenomenon can also be observed with acorns at a time when they are lying thick in the roads. If they are kicked, in any way whatever, they almost invariably skid along for a little way and then rise up on their longest axis, in which position they remain spinning violently, though otherwise stationary on the road. What seems at the present time to be a favourite form of whip-top with children in the streets is an instance of the same thing. It is, roughly, in the shape of a large acorn with a thick cap to it. In whatever position it is first put down on the road violent lashing will always make it rise on to its longest axis, and continued lashing causes it to continue to spin in that position.

Celts. Figs. XI., XII., XIII. (see Plate I.) represent three celts, stone implements used in times past by primitive man for cutting or cleaving, and now discovered from time to time in ancient barrows (Latin: *Celtis*, a chisel).

They provide typical instances of the curious phenomena exhibited by all smooth celts when spun on a smooth horizontal surface (such as a sheet of glass).

Let us perform the same experiment on each in turn and note the results.

In each case, the stone being considered at rest, GAB is a horizontal section (very roughly elliptical) through the centre of gravity G, and GA, GB, GC are mutually at right angles.

FIG. XI. (See end of book.)

(*a*) If a vertical downward tap be applied at A the stone will rock about GB, and if the tap be applied at B the rocking will be about GA.

(*b*) If spun about the vertical GC the stone will continue to spin steadily about GC until brought to rest by friction.

These results are what we should ordinarily expect. But now let us consider the other two celts.

FIG. XII.

(*a*) When a vertical downward tap is applied through A the stone oscillates and then begins, almost immediately, to *rotate* from A to B; but if the tap be applied at B, the stone, after a moment's oscillation, begins to rotate from B to A.

(*b*) If spun in either direction it will begin to oscillate and gradually reverse its direction of spin.

FIG. XIII.

(*a*) A vertical downward tap through A *or* B sets the stone rotating in the direction from A to B.

(*b*) If spun in the direction A to B it will continue to spin steadily in that direction; but if spun in the direction B to A the stone will oscillate violently and eventually reverse its direction of spin, *i.e.* will spin in the direction A to B.

The "Spider tops," which are frequently sold in the streets of London, consist of a heavy little disc mounted on a spindle (Fig. XIV.). When the disc has been set spinning a small curved piece of metal is placed to touch the toe, and at once begins to slide round it, first the side (*a*) in the figure, and then the side (*b*), the motion continuing backwards and forwards till the top comes to rest. The fact is that the toe is magnetic, and this being the case it is easy to see that the rolling of the toe on the side of the metal produces the motion. But Figs. XV. (*a*) and (*b*) illustrate a top, whose spindle behaves in exactly the same manner as the toe of the spider top, and yet *is in no way magnetic.* The action is purely mechanical, as we shall explain in later pages. It is an attractive top to watch, especially as it rushes round the corner when it comes to the end of the coil.*

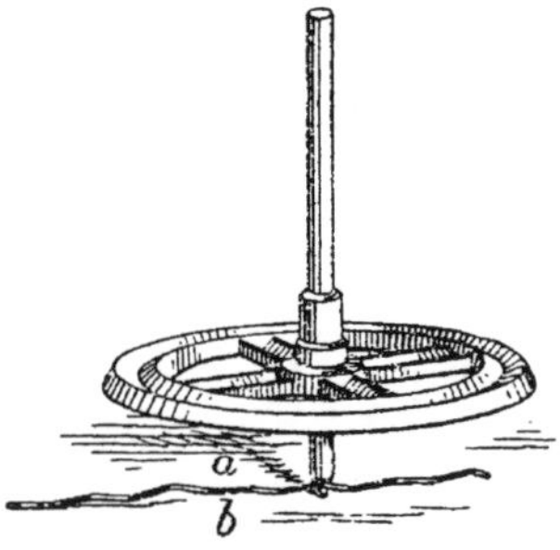

FIG. XIV.

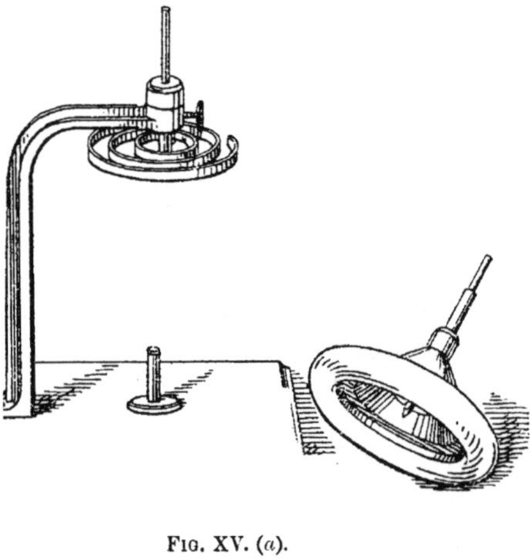
FIG. XV. (*a*).

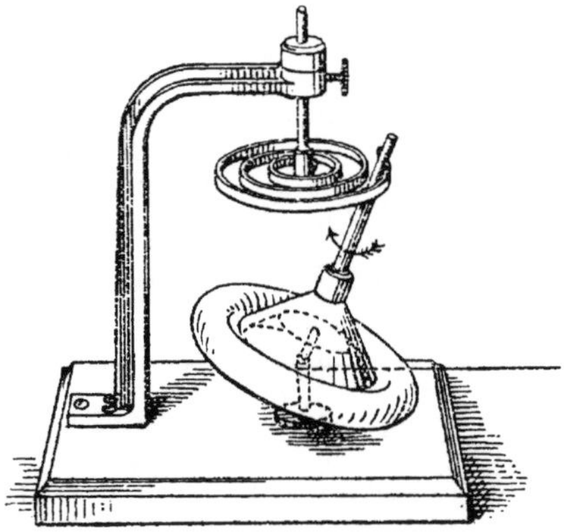
FIG. XV. (*b*).

But the most interesting top of all is undoubtedly the ordinary gyroscope. That depicted in Figs. XVI. to XXI., although merely sold as a toy, is nevertheless capable of illustrating the gyroscopic phenomena which have been so much made use of in modern mechanical invention. If the gyroscope is spun as in Fig. XVII. the surface on which it is standing should be as *smooth* as possible. See page 49, question 14.

---

*This top, which is known as a "Gyroscopic Top," can be purchased of Newton & Co., Scientific Instrument Makers, 3 Fleet Street, London, E.C., price, £1. 6*s*.

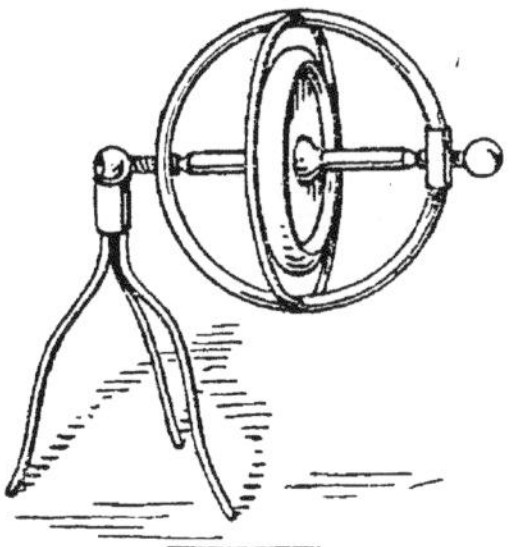

Fig. XVI.

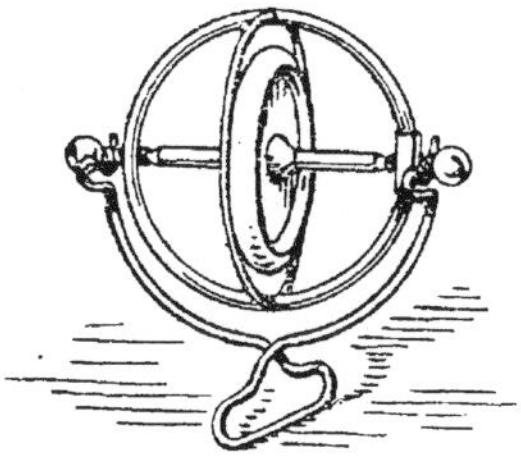

Fig. XVII.

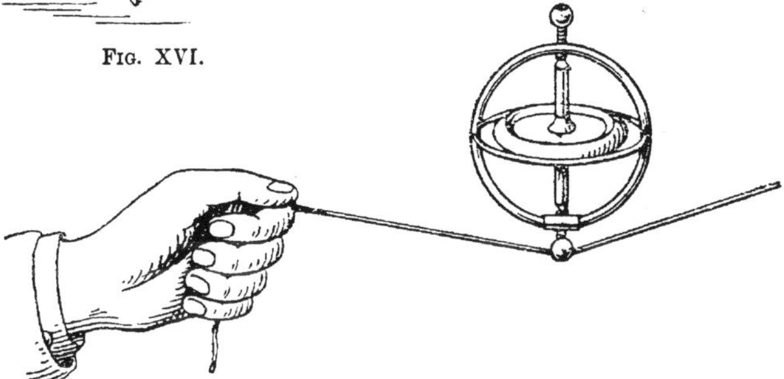

Fig. XVIII.

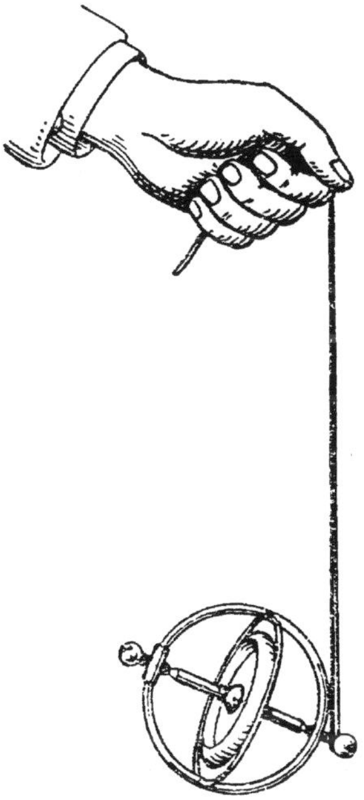

Fig. XIX.

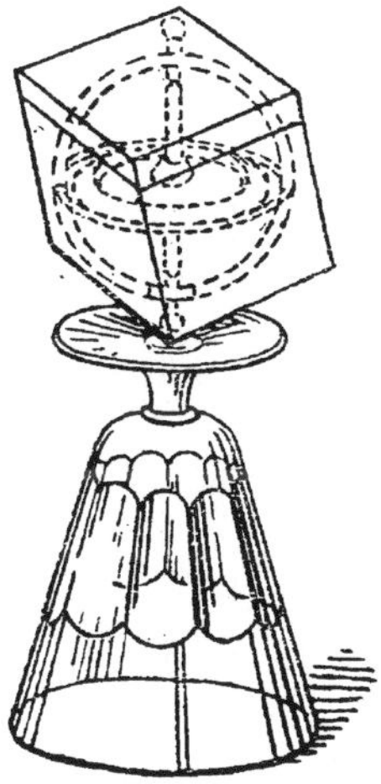

Fig. XX.

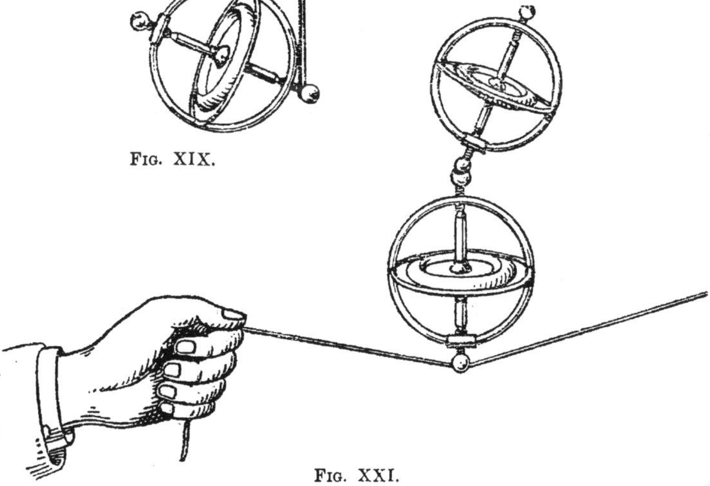

Fig. XXI.

Scientifically constructed Gyroscope. Fig. XXII. represents a scientifically made instrument.* The wheel AB which has (for its size) a very heavy rim, is free to spin about its axle XX', while the frame in which this axle is mounted can turn about the axis YY'. Finally, the frame in which YY' is mounted can turn about the vertical pedestal ZZ', but at Z there is a screw which can be tightened to prevent this turning.

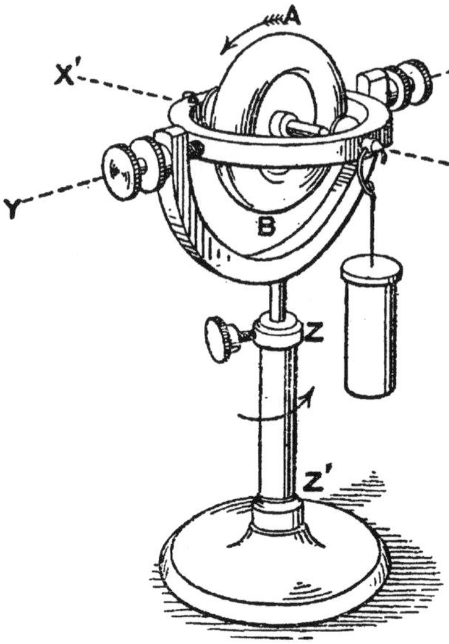

FIG. XXII.

Experiment 1. Suppose now that the wheel is free to turn about all three axes, and that friction is so slight as to be negligible. Let a good spin be given to the wheel in the direction marked, and let a weight be suspended at X. It will be found that, instead of the wheel tilting about $Y'Y$, it will turn about ZZ' in the direction indicated. But if the screw at Z is tightened the revolving wheel will at once turn over about $Y'Y$.

If, when the wheel is free to turn about all three axes, we apply at X a force which is more easily removed than the weight, such as the pressure of a finger, it will be found that, when the rate of spin is small, the wheel will tilt about $Y'Y$ appreciably, and on the finger being removed the whole system will oscillate violently. If, however, the rate of spin is large, only a very small tilting, if any, will be appreciable, while on the removal of the finger the violent oscillations observed previously are now hardly more than a "shiver" of the axle.

* The gyroscope here depicted is known as Wheatstone's Compound, and can be procured of Newton & Co., at the cost of 3 guineas.

Experiment 2. If the wheel is taken out of its bearings at Y, Y', and the framework held in the hands at X, X', a violent attempt to tilt it about $Y'Y$ will, unless proper care is taken, result in the wheel wriggling out of the hands altogether—for the same reason that when properly mounted just now it turned about ZZ'.

It is clear that an opposite direction of spin or any attempt to tilt in the opposite direction, produces an opposite direction of turning about ZZ'.

Experiment 3. Let us now take the gyroscope when the wheel is not spinning, and its axle XX' is inclined at an angle to the vertical as in Fig. XXIII., the screw at Z not being clamped.

If the pedestal ZZ' is held in the hand and slowly swung round, it will be found that XX' turns round about the vertical and points in the same direction as the hand. But let a spin be given to the wheel, and it will be seen at once that, though the hand revolves round the body, XX' *remains always pointing in the same direction in which it pointed originally.**

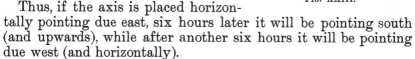

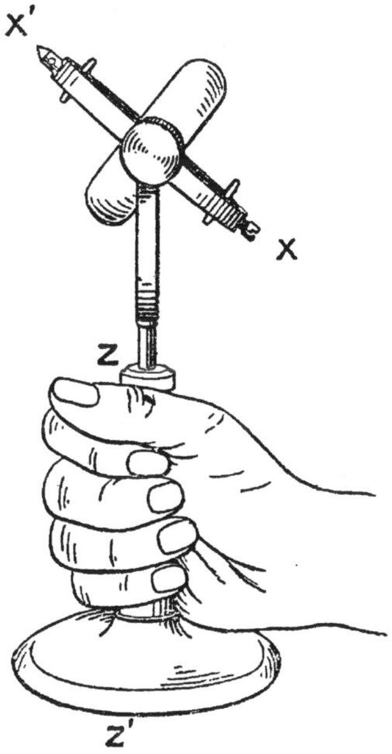

FIG. XXIII.

This stability of the axle of the gyroscope may be employed to prove the rotation of the earth. For suppose the spin is maintained during a period of several hours by means of electric power, and that the gyroscope is set down in a room with its axle pointing to some particular object in the room. After a few hours it will be found that the axle no longer points to that object; showing, not that the axle has changed its position in space, but that the room is in a different position in space owing to the rotation of the earth.

Thus, if the axis is placed horizontally pointing due east, six hours later it will be pointing south (and upwards), while after another six hours it will be pointing due west (and horizontally).

Experiment 4. Now, while the wheel is still spinning, let the turning movement be given to the hand when the screw at Z has been made fast. The gyroscope at once sets itself with the

*The friction may not be quite negligible though very nearly so. The frictional couple at Z is probably sufficiently large to turn the frame YY' round when the wheel is not spinning, but totally inadequate when it is spinning. So much as is appreciable tends to turn it about $Y'Y$, as will be seen later.

axis XX' vertical as in Fig. XXIV., and when once this position has been gained a further swing of the hand in the same direction finds it in stable equilibrium: but if we check or reverse the swing we find that the equilibrium is unstable, and if the checking or reversal be sudden, the gyroscope will turn a complete somersault!

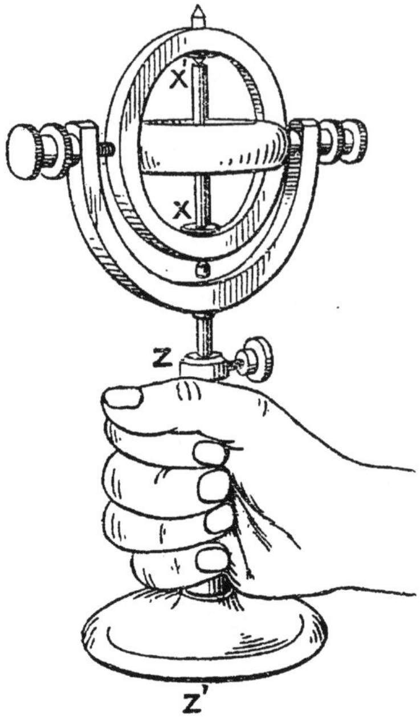

FIG. XXIV.

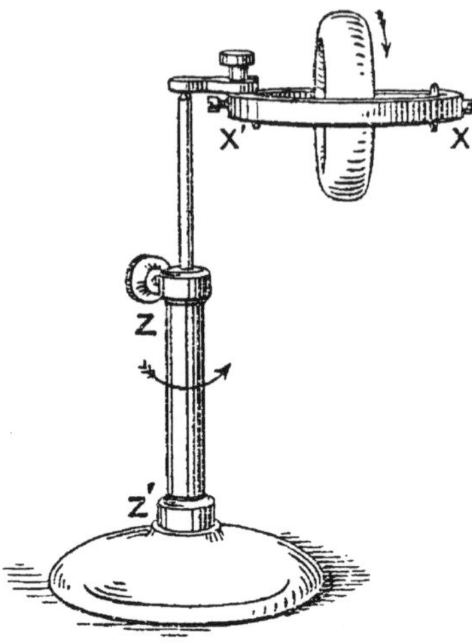

FIG. XXV.

Fig. XXV. represents the same gyroscope taken to pieces and put together in another form. Here the weight of the instrument itself, when no extra weight is suspended, instead of tilting the gyroscope over, causes it to turn about ZZ' as before; and it will be found that if the rate of turning be hurried by the application of a horizontal force to the frame of the gyroscope, then the centre of gravity of the instrument will at once *begin to rise.*

The following pages will be devoted to discussing the phenomena described in this chapter, and to explaining why the gyroscope appears to defy all the recognized laws of gravity. The same principles account for an elongated bullet, when fired from a rifle, always "drifting" to the left-hand side, as also in part for the gyrations of a boomerang, but not in general for the swerving of a sphere, as for instance a cricket ball or a "sliced" golf ball. In order to get a clear idea of the subject it will first be necessary to master a few elementary principles in Dynamics which are discussed in the first two chapters of this book.

CHAPTER I.

ROTATION ABOUT A FIXED AXIS.

1. Definition of angular velocity. If a point P is moving relatively to a point O, but not along the line OP, it is said to have angular velocity about O. The angular velocity is measured by the rate at which the line OP is describing an angle.

For instance (Fig. 1), if O is the centre of a circle round which a point P is moving uniformly with velocity v, having started from the position A, then the rate at which the angle POA is being described is the angular velocity, ω say, of the particle P about O.

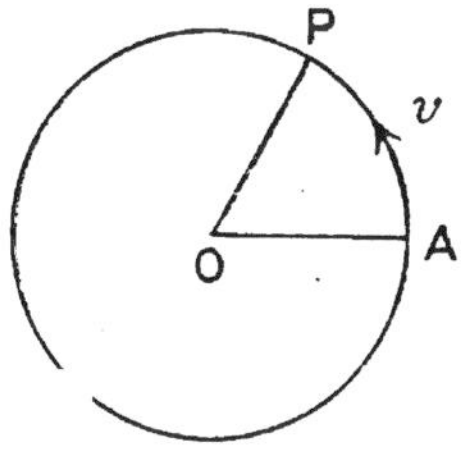

FIG. 1.

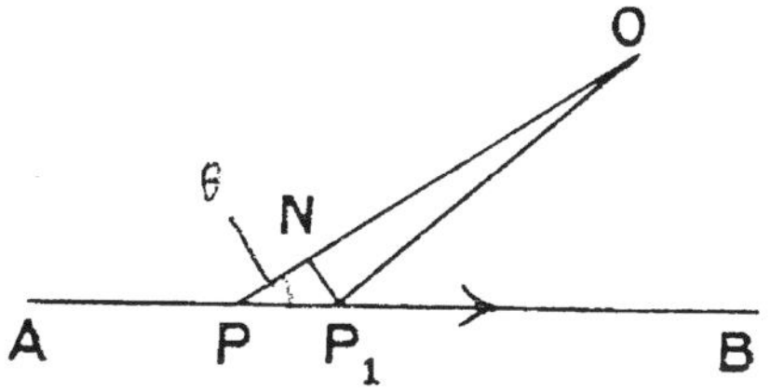

FIG. 2.

In this case if r is the radius of the circle, and v units of length are traversed in a unit of time, say one sec., then $\omega = \frac{v}{r}$ radians per second.

It is clear that when we speak of the angular velocity of a point (P) about a point (O) we really mean the rate at which a *line* (OP) is revolving.

As a further illustration (Fig. 2), suppose P is moving in a *straight line* with velocity v, in the direction AB, and that O is a fixed point.

Let P be the position at the time t, and P_1 at the time $t+\delta t$.

In the small time δt P describes an angle POP_1 about O, and if P_1N is perpendicular to OP, the circular measure of this angle is

$$\frac{P_1N}{OP} = \frac{PP_1 \sin\theta}{OP} = \frac{v\delta t \sin\theta}{OP}; \quad (O\hat{P}B = \theta)$$

· the rate at which this angle is described is $\omega = \frac{v \sin \theta}{OP}$.

∴ the angular velocity of P about O is $\frac{v \sin \theta}{OP}$.

2. The angular velocity of a line in a plane is the rate of increase of the angle it makes with some fixed line in that plane.

For instance, if a stick is so thrown that its motion is in a vertical plane, its angular velocity is the rate of increase of the angle it makes with some vertical line in the plane of motion (Fig. 3).

Notice that we speak of the angular velocity of a point *about a point*, but of the angular velocity "of a line," simply—not "about" anything.

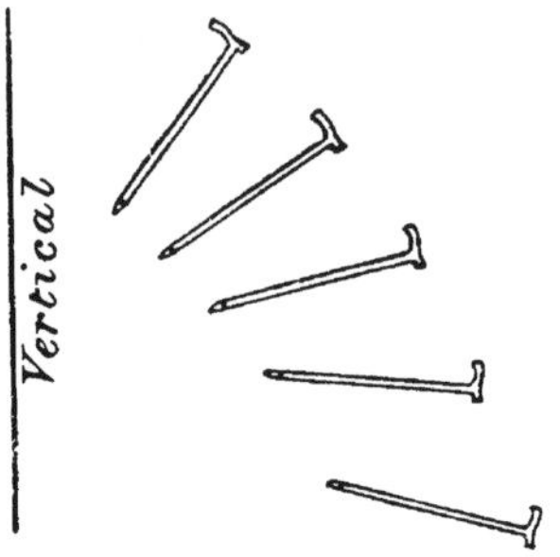

FIG. 3.

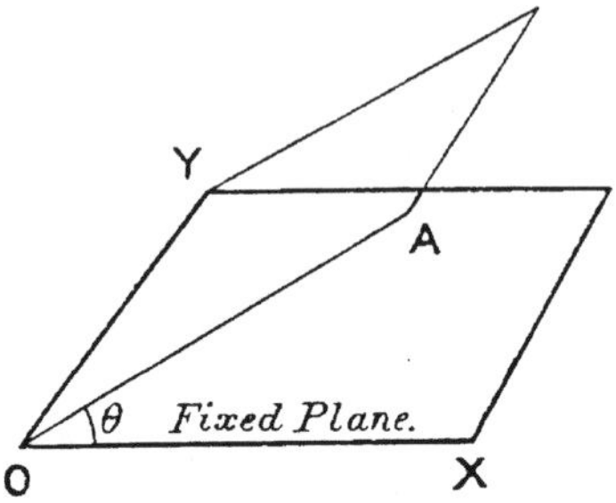

FIG. 4.

3. Angular velocity of a plane. The angular velocity of a plane is the rate of increase of the angle it makes with any plane *fixed in space*.

For instance, in Fig. 4, if θ is at any instant the angle between the fixed and revolving planes, the angular velocity, ω say, is $\frac{d\theta}{dt}$.

4. Rigid Body. By a rigid body or system we mean one in which the particles composing the body or system never alter their positions relative to each other.

An ordinary walking stick is an approximate instance of a rigid body: an umbrella is not.

5. Angular velocity of a rigid body. The angular velocity of a rigid body is the angular velocity of any plane fixed in the body.

In the accompanying figure the body is represented by a cylinder. If θ is at any instant the angle between the two planes, then the angular velocity of the cylinder is

$$\omega = \frac{d\theta}{dt}.$$

If the plane YOX were not fixed in space, but revolving about OY, then ω, if measured by $\frac{d\theta}{dt}$, would be the angular velocity of the body *relative to the plane* YOX and not "the angular velocity of the body."

Angular acceleration is the rate of increase of angular velocity.

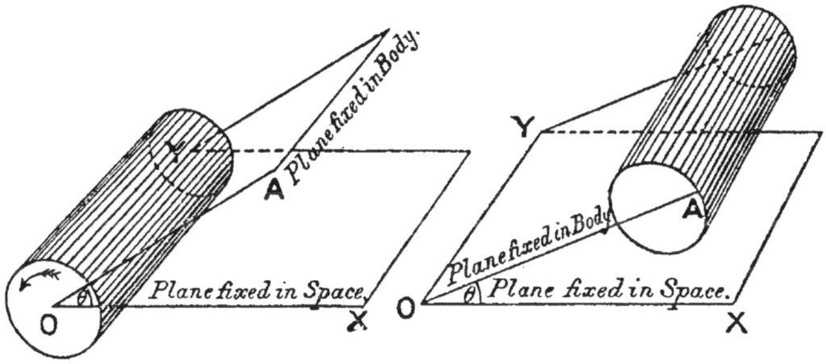

FIG. 5.

6. Torque. A force, or system of forces, capable of turning a body about any axis is said to be, or to have, a torque about that axis. It is measured by the moment of the forces about the axis.

7. Uniform angular acceleration. If a rigid body is rotating about a fixed axis with *uniform* angular acceleration α, then after a time t it is clear that, supposing the initial velocity to have been ω, the final velocity

$$\Omega = \omega + \alpha t. \qquad \text{.........................(1)}$$

It is also clear to those familiar with the elementary formulae of particle dynamics, that if in time t, the rigid body under the same conditions turns through an angle θ, the *average* velocity at which this angle is described is

$$\frac{\Omega + \omega}{2};$$

$$\therefore \theta = \frac{\Omega + \omega}{2} \cdot t$$

$$\theta = \frac{\omega + \alpha t + \omega}{2} \cdot t$$

$$\theta = \omega t + \tfrac{1}{2}\alpha t^2. \qquad \text{.....................(2)}$$

Combining equations (1) and (2) as in particle dynamics we obtain

$$\Omega^2 = \omega^2 + 2\alpha\theta. \qquad \text{........................(3)}$$

These formulae are exactly analogous to the similar ones relating to the rectilinear motion of a particle under uniform rectilinear acceleration; for they are obtained from exactly similar considerations; but the student may feel at first that he is dealing with a different *kind* of velocity, and a different (*i.e.* angular) measurement of space. In order, therefore, to obtain a clearer realisation of the quantities employed, the *dimensions* of the equations should be considered, this being a most important method of checking algebraical expressions.

Dimensions. (1) $\Omega = \omega + at;$

$$\frac{1}{[T]} = \frac{1}{[T]} + \frac{1}{[T]^2} \times [T],$$

$$[T]^{-1} = [T]^{-1} + [T]^{-1}$$

(2) $\theta = \omega t + \frac{1}{2}at^2;$

$$[T]^0 = \frac{[T]}{[T]} + \frac{[T]^2}{[T]^2}.$$

(3) $\Omega^2 = \omega^2 + 2a\theta;$

$$[T]^{-2} = [T]^{-2} + [T]^{-2}.$$

EXAMPLES.

1. A wheel acquires a velocity of 100 rad./sec. in 16 secs. under a uniform angular acceleration, having started from rest. Find the acceleration.

2. Through what angle has it revolved in the 16 secs.? What angle will it turn through in the next 16 secs.?

3. A wheel is given an initial velocity of 1,000 revolutions per sec., after which it is acted on by a uniform retardation which in 3 secs. reduces its velocity by 12 revolutions a second. How many revolutions will it have made from the start when it is making 20 revolutions a second?

8. Inertia. Newton's First Law of Motion asserts that, *Every body will continue in its state of rest or of uniform motion in a straight line, except in so far as it is compelled by impressed forces to change that state*; which is equivalent to saying that a body has no power of itself to change its state of rest or of uniform motion in a straight line. This statement is sometimes called the "Law of Inertia," "inertia" being regarded as a *property* of mass (Latin: *iners*).

Newton's Third Law of Motion asserts that, *To every action there is an equal and opposite reaction*; and thus the existence of a force implies of necessity some resistance against which the force is applied. When a force is applied to a body at rest the resistance offered by the body is due to the inertia of its mass. If the body had no mass no force could be exerted on it. So also, if we attempt to alter the *existing* motion of a body, a resistance is experienced. This co-existence of action and reaction is described by the term *stress*; and the resistance which a body offers by reason of its mass is sometimes called the *force of inertia* of the body, meaning the resistance *due to* inertia.

9. Let us now imagine a rigid body to be rotating about a fixed axis AB with angular velocity ω, as for instance a gate swinging on its hinges, Fig. 6.

Any given particle, such as P, in the body is moving with a definite linear velocity v, perpendicular to AB, which is represented by $r\omega$, r being the length of the perpendicular from the particle in question upon the axis.

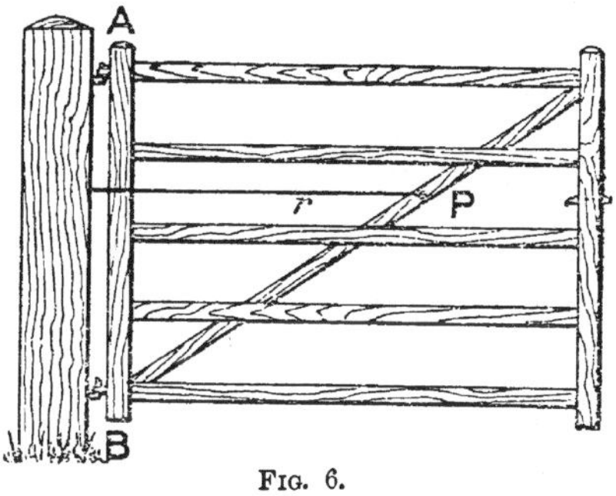

FIG. 6.

Therefore if m is the mass of the particle, its linear momentum is mv or $mr\omega$: and if we take the "moment" of this momentum about AB, as we take the "moment" of a force, then the "moment of momentum" of the particle about AB is

$$mv \,.\, r = mr\omega \,.\, r = mr^2\omega\,;$$

and the "moment of momentum" of the whole body is $\Sigma mr^2\omega$, or $\omega\Sigma mr^2$ since ω is the same for all the particles.

Now Σmr^2 is a quantity dependent in each case on the configuration of the particles forming the body, and on the position of the axis AB; and it is clear that for any particular body and axis an expression Mk^2 can be found equal to Σmr^2, where M is the mass of the whole body and k some length which is constant for the given body and axis. This expression is termed the "moment of inertia" of the body (see Art. 10) about the axis in question, and is frequently denoted by I. It is a most important quantity whenever the motion of a rotating body is considered; for the rate of change of $\Sigma mr^2\omega$ or $Mk^2\omega$, *i.e.* the rate of change of the moment of the momentum, represents, as we shall see later, the turning effect about the axis in question of the external acting forces, and therefore enables us to determine the motion which these forces cause.

The length k is called the *radius of gyration* of the body about the axis.

The moment of the momentum of a body about an axis, *i.e.* $Mk^2\omega$, is frequently called the *angular momentum* of the body about that axis.

QUESTIONS.

[**1.**] What are the dimensions of

(α) Moment of inertia?

(β) Angular momentum?

(γ) The moment of a force, *i.e.* a "torque"?

[**2.**] A heavy iron door weighs a cwt. Taking its radius of gyration as 2 ft., what is its moment of inertia about its hinges? State clearly the units you employ.

10. The phrase "*moment* of inertia" means in the first instance "the importance of the inertia"; and its significance will be easily grasped if we consider the obvious importance in the motion of some rigid bodies of the configuration or arrangement of the particles which form them.

A golfer might conceivably possess a "putter" of the same total mass as his "driver"; but the difference in their usefulness for conveying force to the ball is clear at once. This is of course due to the different configuration of particles in the two clubs. Similarly, a cricketer "takes the long handle" when hard hitting is his primary object.

Or again, suppose we have two wheels, of the same total mass, rotating about fixed axes with the same angular velocity, the only difference in them being that one has most of its mass concentrated near the centre, the other at the rim. It will be found that the one with the heavy rim requires a much greater force to stop its rotation than the one with a heavy centre and light rim.

The devices for finding the value of different moments of inertia, either without or with the aid of the Calculus are discussed fully in various text books. The following two elementary examples will illustrate the method when the Calculus is employed.

11. *To find the moment of inertia of a thin rod of mass M and length l, about an axis which passes through one end of the rod and is perpendicular to its length.*

Let OA be the rod of which PQ is a small element dx, at a distance OP $(=x)$ from the axis, Fig. 7.

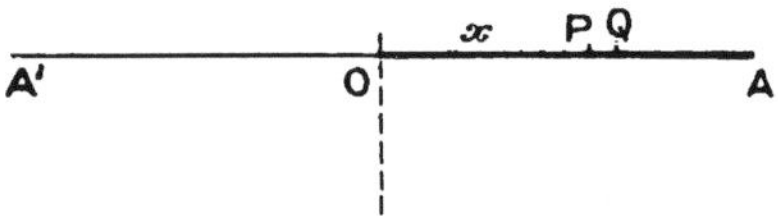

FIG. 7.

Let μ = mass per unit length of the rod. Then the moment of inertia of the whole rod

$$= \Sigma \mu \, dx \, . \, x^2$$

$$= \int_0^l \mu \, dx \, . \, x^2$$

$$= \left[\frac{\mu x^3}{3} \right]_0^l$$

$$= \mu \frac{l^3}{3}$$

$$= M \frac{l^2}{3},$$

$$k^2 = \frac{l^2}{3}, \quad \textit{l being the length of the rod.}$$

From the definition of moment of inertia as being the value of Σmr^2, it is clear that the moment of inertia for any body is the sum of the moments of inertia of its component parts, and, therefore, if O were the middle point of a rod $A'A$, the moment of inertia about the central perpendicular axis would be $2\frac{M'l^2}{3}$ (M' being the mass of a length l which is in this case *half* the rod), or $\frac{\mathbf{Mh^2}}{3}$,

M being the whole mass and h half the length of the rod.

To find the moment of inertia of a solid disc of mass M, radius a, and thickness b, about its central perpendicular axis.

Let us consider first a thin circular section ABC of the disc (Fig. 8), the whole mass of which is dM, and let μ represent the mass of this section per unit area. Divide the section further into thin rings, one of which is represented in the figure of radius r and thickness dr.

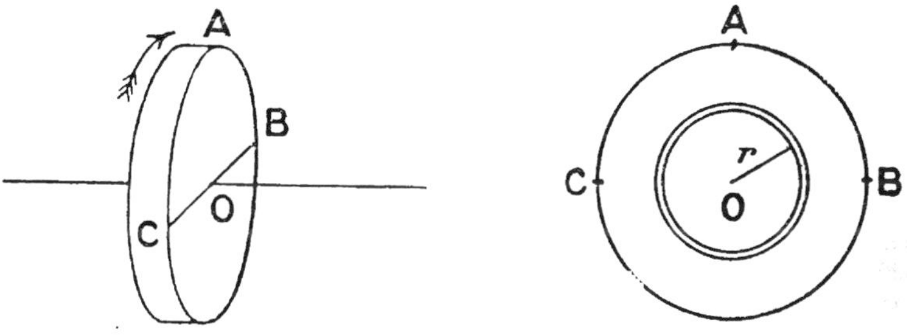

FIG. 8.

The mass of this ring is $\mu 2\pi r\,dr$.

Every particle of the ring is at a distance r from the axis;

the moment of inertia of the ring is

$$\mu 2\pi r\,dr\,.\,r^2, \text{ or } 2\mu\pi r^3 dr;$$

∴ the moment of inertia of the thin section ABC

$$= \int_0^a 2\mu\pi r^3 dr$$

$$= 2\left[\frac{\mu\pi r^4}{4}\right]_0^a$$

$$= \mu\frac{\pi a^4}{2}$$

$$= \frac{dM}{2}a^2,$$

where dM is the mass of the thin circular section.

Now if we consider the (thick) disc to be made up of an infinite number of thin sections exactly similar to ABC, each having a moment of inertia $\frac{dM.a^2}{2}$ about the axis in question, then for the whole solid disc made up of an infinite number of particles each of mass m, the moment of inertia

$$\Sigma mr^2 \text{ is equal to } \int dM\frac{a^2}{2}=\frac{Ma^2}{2},$$

$$\text{and} \quad k^2=\frac{a^2}{2}, \text{ i.e. the radius of gyration is } \frac{a}{\sqrt{2}}.$$

[3] How has the thickness b come in?

[4.] What is the angular momentum ($Mk^2\omega$) of a grindstone about its axis, in pound-foot-radian-second units, if its mass is 20 lbs., radius 1 ft., and it is making two revolutions a second?

A list of moments of inertia for some simple solids will be found at the end of this chapter.

12. Rate of change of angular momentum. It was stated in Art. 9 that the rate of change of the angular momentum ($I\omega$) of a rigid body rotating about a fixed axis represents the turning effect about the axis in question of the external acting forces: *i.e.* the moment of the forces about the axis.

Let us consider a body turning about a fixed axis (through O perpendicular to the plane of the paper in Fig. 9), under forces whose moment about the axis is K, with angular velocity ω, and angular acceleration α. Let P be any particle of mass m,

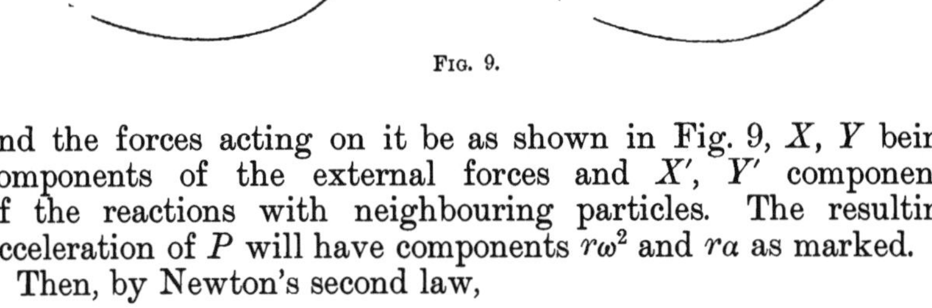

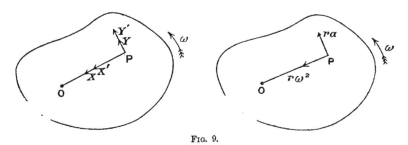

FIG. 9.

and the forces acting on it be as shown in Fig. 9, X, Y being components of the external forces and X', Y' components of the reactions with neighbouring particles. The resulting acceleration of P will have components $r\omega^2$ and $r\alpha$ as marked.

Then, by Newton's second law,

$$Y+Y'=mr\alpha\cdot$$

$$(Y+Y')r=mr^2\alpha.$$

Taking then the motion of all the particles,

$$\Sigma(Y+Y')r=\Sigma mr^2\alpha$$
$$=\alpha\Sigma mr^2,$$

since α is the same for all the particles,

$$=I\alpha.$$

But since the internal reactions are equal and opposite, their moment about the axis is zero.

$$\therefore\ \Sigma Y'r \text{ is zero};$$
$$\therefore\ \Sigma Yr=I\alpha.$$

But ΣYr represents the moment of the external forces about the axis, *i.e.* K;

$$\therefore\ K=I\alpha,$$

that is, **about any fixed axis, the moment of the external forces is equal to the rate of change of the angular momentum.**

Dimensions. $\dfrac{[M][L]}{[T]^2}\cdot[L]=[M][L]^2\cdot\dfrac{1}{[T]^2}.$

13. The student should notice that as in linear motion we have the equation

$$\text{force}=\text{inertia}\times\text{acceleration},$$

so in rotational motion we have

$$\text{moment of force}=\text{moment of inertia}\times\text{angular acceleration}.$$

A similar analogy *of form* can be seen in all equations relating to rotational motion.

14. Change of angular momentum. From the above equation we see that if α and K are constant,

$$Kt=I\alpha\,.\,t$$
$$=I(\omega_1-\omega),$$

where ω becomes ω_1 after time t.

Further, this is true when K and α vary, if t is so small that they may be regarded as constant during the interval.

If now K be made infinite and t indefinitely small in such a way that Kt is finite and equal to K', we have an impulsive couple of moment K' and

$$\therefore\ K'=I(\omega_1-\omega),$$

that is, **about any fixed axis, the moment of the external impulse is equal to the change of the angular momentum.**

[5.] What is the analogous equation in linear motion?

15. Kinetic Energy. *If a body is rotating with angular velocity ω about a fixed axis its kinetic energy is $\frac{1}{2}I\omega^2$.*

For, using the notation of Art. 12, the linear velocity v of P is $r\omega$.

∴ kinetic energy of particle $m = \frac{1}{2}mv^2 = \frac{1}{2}mr^2\omega^2$

„ whole body $= \Sigma\frac{1}{2}mr^2\omega^2$

$$= \frac{\omega^2}{2}\Sigma mr^2$$

$$= \tfrac{1}{2}Mk^2\omega^2,$$

or $\frac{1}{2}I\omega^2$

Dimensions. $$\text{K.E.} = \tfrac{1}{2}I\omega^2,$$

or $$\frac{[M][L]^2}{[T]^2},$$

i.e. same dimensions as force × distance, which is work done.

[6.] What is the analogous equation in linear motion?

16. Work done by a couple. *If a constant couple K turns a rigid body about a fixed axis through an angle θ, then the work done by the couple on the body is Kθ.*

For, in the small element of time δt, let the body turn through a small angle $\delta\theta$.

Then P (see Fig. 9) is displaced through a distance $r\delta\theta$ along the direction of the force $Y + Y'$.

∴ The work done on the particle P in time δt

$$= (Y + Y')r\delta\theta.$$

∴ The work done on the body in time δt

$$= \Sigma(Y + Y')r\delta\theta$$
$$= \delta\theta\Sigma(Y + Y')r$$
$$= \delta\theta\Sigma Yr, \text{ since } \Sigma Y'r = 0$$
$$= \delta\theta \,.\, K,$$

where K is the moment of the couple.

Hence, the couple K being constant, the work done by it during the whole time is given by $K\theta$.

Since the work done on any body is equal to the change in kinetic energy, we have

$$K\theta = \tfrac{1}{2}I(\omega_1^2 - \omega^2),$$

if ω changes to ω_1.

This also follows from the equations

$$K = I\alpha, \quad \text{Art. 12},$$

and $$\omega_1^2 = \omega^2 + 2\alpha\theta, \quad \text{Art. 7},$$

whence $$\omega_1^2 - \omega^2 = \frac{2\theta K}{I}$$

or $$K\theta = \tfrac{1}{2}I(\omega_1^2 - \omega^2).$$

Dimensions. $$\frac{[M][L]}{[T]^2}[L] = [M][L]^2 \cdot \frac{1}{[T]^2}.$$

[7.] What is the analogous equation in linear motion?

17. Summary of preceding results.

Linear Motion.		Rotational Motion.	
Inertia or mass	$=M$.	Moment of inertia	$=Mk^2=I$.
Momentum	$=Mv$.	Moment of Momentum or Angular Momentum	$=I\omega$.
Force	$=Mf$.	Moment of force	$=I\alpha$.
Impulse	$=M(v_1-v)$.	Moment of impulse	$=I(\omega_1-\omega)$.
Kinetic energy	$=\frac{1}{2}Mv^2$.	Kinetic energy	$=\frac{1}{2}I\omega^2$.
Work done by constant force P moving its point of application through distance s	$=Ps=\frac{1}{2}M(v_1^2-v^2)$.	Work done by constant couple K in turning body through angle θ	$=K\theta=\frac{1}{2}I(\omega_1^2-\omega^2)$.

18. Table of moments of inertia for some simple solids of common occurrence. Most of the following results (the exceptions are noted by an asterisk) are included in a rule enunciated originally by the late Dr. Routh, and known as "Routh's Rule."

"The moment of inertia of a solid body about an axis of symmetry $= \text{mass} \times \dfrac{\text{sum of squares of perpendicular semi-axes}}{3,\ 4,\ 5}$, the denominator being 3, 4, or 5 according as the body is rectangular, elliptical, or ellipsoidal."

It should be remarked that these perpendicular axes are axes of symmetry, and that a circle and sphere are special cases of an ellipse and ellipsoid respectively. The results can be obtained by simple integration similar to that indicated in the earlier part of the chapter. *Except where otherwise stated the axes are taken through the centre of gravity.*

(1) Thin straight rod—about an axis perpendicular to its length:

* Through one end, $\dfrac{Ml^2}{3}$, where l is the length.

Through the centre of gravity, $\dfrac{Mh^2}{3}$, where h is the half-length.

(2) Thin rectangular lamina, sides $2a$, $2b$—about an axis:

Perpendicular to plane, $M.\dfrac{a^2+b^2}{3}$.

Parallel to side $2a$, $\dfrac{Mb^2}{3}$.

(3) Rectangular parallelopiped sides $2a$, $2b$, $2c$:

Perpendicular to $2b$, $2c$, $M.\dfrac{b^2+c^2}{3}$.

(4) Circular wire, radius a:

* Perpendicular to plane, Ma^2.

* Diameter, $\frac{Ma^2}{2}$.

(5) Circular disc, radius a:

Perpendicular to plane, $\frac{Ma^2}{2}$.

Diameter, $\frac{Ma^2}{4}$.

(6) Circular cylinder, radius a:

Central axis, $\frac{Ma^2}{2}$.

(7) Elliptical disc, axes $2a$, $2b$ ·

Axis $2a$, $\frac{Mb^2}{4}$.

Axis $2b$, $\frac{Ma^2}{4}$.

Perpendicular to plane, $M.\frac{a^2+b^2}{4}$.

(8) Spherical shell, radius a:

* Diameter, $\frac{2Ma^2}{3}$.

(9) Sphere:

Diameter, $\frac{2Ma^2}{5}$.

(10) Ellipsoid, axes $2a$, $2b$, $2c$:

Axis $2a$, $M.\frac{b^2+c^2}{5}$, and similarly for other axes.

(11) * Right cone, height h, radius of base a:

Axis, $M.\frac{3a^2}{10}$.

Perpendicular to axis, $M.\frac{3}{20}\left(a^2+\frac{h^2}{4}\right)$.

19. We will now proceed to the discussion of some problems in connection with the angular momentum and kinetic energy of a rigid body rotating about a fixed axis, on the assumption that the general principles of work and energy are understood.

ILLUSTRATIVE EXAMPLES.

1. A spinning top weighing 8 oz. is set rotating with a velocity of 5 revolutions a second. Taking the radius of gyration as 2 ins., calculate (i) its angular momentum ; (ii) its kinetic energy.

(i) Angular momentum $=I\omega$.

Here $\omega=5 \cdot 2\pi=10\pi$ radians a second.

$$I=\frac{1}{2}\left(\frac{2}{12}\right)^2=\frac{1}{2}\frac{1}{36} \text{ lb.-ft.}^2.$$

Angular momentum $=I\omega$

$$=\frac{1}{2}\frac{1}{36} \cdot 10\pi$$

$$=\frac{5}{36}\pi \text{ lb.-ft.}^2\text{/sec.}$$

(ii) Kinetic energy $=\frac{1}{2}I\omega^2$

$$=\frac{1}{2} \cdot \frac{1}{2} \cdot \frac{1}{36} 100\pi^2 \quad [\pi^2=10 \text{ approx.}]$$

$$=\frac{250}{36}=\frac{125}{18}=7 \text{ ft.-pdls. approx.}$$

2. Round the spindle of a top $\frac{1}{2}$ in. in diameter, is wound a thin string 3 ft. in length which is pulled off by a uniform force in 2 secs.

(i) Find the angular velocity after that time.

(ii) If the angular velocity is 15π rad./sec., how long did it take to pull off the string?

(i) The point where the string is held passes over 3 ft. in 2 secs.

$\therefore$ its *average* velocity $=\frac{3}{2}$ ft./sec.

$\therefore$ since the pulling force is uniform its final velocity $=3$ ft./sec. (Art. 7), and this is the final velocity of any point on the spindle's rim.

Hence, since the spindle is $\frac{1}{4}$ in. in radius or $\frac{1}{48}$ ft., the final angular velocity of the spindle or of the top $=3\div\frac{1}{48}=144$ rad./sec.

(ii) Let t secs. be the unknown time.

Then as before, the final velocity of a point on the string or on the rim of the spindle

$$=2 \cdot \frac{3}{t} \text{ ft./sec.}$$

The final angular velocity of top therefore

$$=\frac{2 \cdot 3}{t}\div\frac{1}{48} \text{ rad./sec.}$$

But this also $=15\pi$ rad./sec.

$$\therefore \frac{6 \cdot 48}{t}=15\pi;$$

$$\therefore t=\frac{6 \cdot 48}{15\pi}=\frac{96}{5\pi} \text{ secs.}$$

3. A heavy top weighing $1\frac{1}{2}$ lbs., whose radius of gyration may be taken as 4 ins., is revolving at the rate of 450 turns a minute.

(i) What energy is expended in stopping it?

(ii) If a frictional force equal to $\frac{5\pi}{96}$ lbs. wt. be applied to the rim at a distance of 6 ins. from the axis, how soon will the top be brought to rest, supposing the axis to be fixed vertically all the time?

(iii) How many turns will it have made since the force was applied?

(i) Here $I=\frac{3}{2}\left(\frac{4}{12}\right)^2$ or $\frac{1}{6}$ lb.-ft.^{2},

$$\omega=\frac{450 \cdot 2\pi}{60} \text{ rad./sec.}=15\pi.$$

The kinetic energy of the top (which must require an equal amount of energy to stop it)

$$=\tfrac{1}{2}I\omega^2$$
$$=\tfrac{1}{2}\,.\,\tfrac{1}{6}\,.\,15^2\,.\,\pi^2 \quad [\pi^2=10 \text{ approx.}]$$
$$=187\tfrac{1}{2} \text{ ft.-pdls.}$$

(ii) We must first find the angular retardation produced.

Denoting this by α rad./sec.^{2} we have

rate of change of angular momentum $=I\alpha$

$$=\frac{1}{6}\alpha \text{ lb.-ft.-rad.-sec. units.}$$

$$\text{Moment of applied torque}=\frac{5\pi}{96}\,.\,32\,.\,\frac{1}{2} \text{ units.}$$

Hence,
$$\frac{5\pi}{96}\,.\,32\,.\,\frac{1}{2}=\frac{1}{6}\alpha\,;$$

$$\therefore\ \alpha=\frac{5\pi}{96}\,.\,32\,.\,3=5\pi\,;$$

$$\therefore\ \text{required time}=\frac{\omega}{\alpha}$$
$$=\frac{15\pi}{5\pi}$$
$$=3 \text{ secs.}$$

Or we may proceed directly.

Change of angular momentum in time t

$$=I\omega=\tfrac{1}{6}15\pi \text{ units.}$$

Moment of applied torque

$$K=\frac{5\pi}{96}\,.\,32\,.\,\frac{1}{2} \text{ units.}$$
$$Kt=I\omega\,;$$
$$\therefore\ t=\frac{15\pi}{6}\div\frac{5\pi}{96}\,.\,16$$
$$=3 \text{ secs. as before.}$$

(iii) Since the final angular velocity $=0$,

and initial „ „ $=15\pi$ rad./sec.;

∴ average „ „ $=\dfrac{15\pi}{2}$ rad./sec.

$$=\frac{15\pi}{2}\div 2\pi \text{ revolutions/sec.}$$
$$=\frac{15}{4}\,;$$

∴ in 3 secs. top will have made $\dfrac{45}{4}$ or $11\tfrac{1}{4}$ revolutions.

Or we may use the equation $K\theta=\tfrac{1}{2}I(\omega_1^2-\omega^2)$.

For from (i) Kinetic energy $=187\tfrac{1}{2}$ ft.-pdls.

$$\text{Applied couple } K=\frac{5\pi}{96}\times 32\times\frac{1}{2} \text{ ft.-pdls.}$$

$\therefore$ if θ be the angle through which G moves in destroying the kinetic energy,

$$\frac{5\pi}{96} \cdot 32 \cdot \frac{1}{2} \cdot \theta = 187\tfrac{1}{2}.$$

$$\theta = \frac{375 \cdot 3}{5\pi} \text{ radians}$$

$$= \frac{75 \cdot 3}{\pi} \cdot \frac{1}{2\pi} \text{ revolutions}$$

$$= \frac{45}{4} \quad (\text{taking } \pi^2 = 10)$$

$$= 11\tfrac{1}{4} \text{ as before.}$$

4. If the mass of the top in example (2) be 8 oz., its radius of gyration about its axis of symmetry 2 ins., and the pulling force be 10 lbs. weight throughout, how many revolutions a second will it be making when all the string is unwound?

The work done $= 10 \cdot 3 \cdot 32$ ft.-pdls.

If ω = the angular velocity generated, in radians per sec.,

$$\text{kinetic energy} = \frac{1}{2} I\omega^2$$

$$= \frac{1}{2}\frac{1}{2}\left(\frac{1}{6}\right)^2 \omega^2.$$

Equating the K.E. to the work done, we get

$$\omega^2 = 10 \cdot 3 \cdot 32 \cdot 4 \cdot 36;$$

$$\therefore \omega = 96\sqrt{15};$$

$$\therefore \text{number of revolutions per sec.} = \frac{48\sqrt{15}}{\pi}.$$

Notice that in this case the size of the spindle does not affect the problem. If the spindle were larger the string would be unwound in fewer turns, *but in a shorter time.* The force would act at a longer arm, and the power or rate of work would be greater, but the actual work done, or kinetic energy generated, would be the same. For the same reason the string need not be "thin."

5. To a wheel and axle of mass 48 lbs. and radius of gyration 6 ins. is attached a weight of 13 lbs. by means of a rope wound round the axle. Taking the radius of the latter as 4 ins., and neglecting the weight of the rope, find the velocity of the 13 lbs. weight after it has descended 52 ft.

We shall equate the work done by gravity on the whole system to the kinetic energy of the system. Let v ft./sec. be the required velocity of the weight: then the corresponding angular velocity of the wheel is

$$v \div \frac{1}{3} = 3v \text{ rad./sec.}$$

The K.E. of the whole system

$$= \frac{1}{2} \cdot 48 \cdot \frac{6^2}{12^2}(3v)^2 + \frac{1}{2} 13 \cdot v^2$$

$$= \frac{v^2}{2} \cdot 121 \text{ ft.-pdls.}$$

Work done by gravity on the system

$$= 13 \cdot 32 \cdot 52 \text{ ft.-pdls.}$$

Hence $\frac{121}{2}v^2=13.32.52,$

$$v^2=\frac{2.13.32.52}{121},$$

$$v=\frac{26.8}{11}$$

$$=19 \text{ ft./sec. nearly.}$$

6. The following example involves the use of the Calculus.

Find the energy communicated to a top when it is set in motion by a string $1\frac{1}{2}$ yds. in length, being pulled with a force which varies as the length of the string already unwound.

Suppose that x ft. have been unwound at any moment. Then the pull at that moment $=\lambda x$ units of force. The whole work done or energy communicated when $1\frac{1}{2}$ yds. are unwound

$$=\lambda\int_0^{4\frac{1}{2}} x\,dx$$

$$=\lambda\frac{x^2}{2}\Big]_0^{4\frac{1}{2}}$$

$$=\frac{81\lambda}{8} \text{ units of energy.}$$

Here neither the size of the spindle nor the radius of gyration make any difference.

EXAMPLES FOR SOLUTION.

1. A spinning top weighing 6 oz. makes 300 revolutions a minute. Taking the radius of gyration about the axis of revolution (supposed vertical) as $1\frac{1}{2}$ ins., calculate

(i) its angular momentum; (ii) its energy.

2. It is said that a Diabolo spool in full rotation spins 2000 turns a minute. Taking the mass as $3\frac{1}{2}$ oz. and the radius of gyration as $\frac{1}{2}$ in., find its angular momentum and how much energy has been expended on it, assuming that half has been wasted.

3. The weights of the minute, hour, and seconds hands of a watch are as 15 : 10 : 1. Compare their angular momenta, their lengths being as 3 : 2 : 1, assuming that the minute and hour hands revolve about one end, but the seconds hand about its middle point.

(For the value of k, see Art. 18.)

4. A door a ft. wide is shutting with angular velocity ω rad./sec. If it comes to rest in $\frac{1}{10}$ sec., find the shock on the door post, the mass of the door being M lbs. and its radius of gyration k ft.

5. A door weighing 40 lbs. is rotating about its hinges, when its edge, 3 ft. 6 ins. away from the hinges, and moving at 30 mi./hr., meets an immovable object. What is the measure of the shock experienced by that object during $\frac{1}{4}$ of a second? Radius of gyration $2\frac{1}{2}$ ft.

6. Given that the top in Fig. II. weighs $3\frac{1}{2}$ oz., that the string is $1\frac{1}{2}$ yds. long, and the average radius of the conical body is 1 in., calculate the angular momentum and the energy due to rotation when the top falls from rest so as to unwind in $1\frac{1}{2}$ secs.

Take the radius of gyration as $\frac{2}{3}$ inch.

7. If the top in the preceding question is making 360 turns a minute just after falling, how long did it take to fall?

8. The following method of spinning a top is frequently employed. A rod AB fitted with a spiral thread can be pressed vertically downwards along the axis of symmetry of the top, thus causing the top to rotate. The rod can be raised up and pressed down again, and so on (Fig. 10).

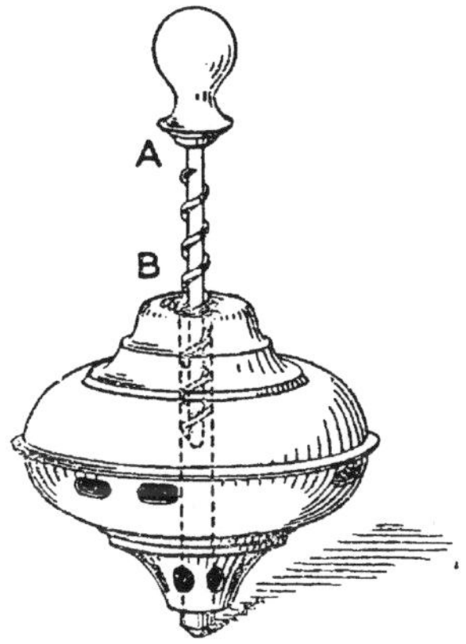

Fig. 10.

Find the energy communicated to the top after pressing down five times with a force of 6 lbs., the rod BC being 6 ins. If the top weighs 5 oz. and the radius of gyration about its axis is 3 ins., what is its angular velocity after the five downward strokes?

9. In the last question how many times must the sleeve be pressed down with a force of 5 poundals to give the top an angular velocity of 32 radians a second?

10. The steering of a torpedo is effected by means of a gyroscope mounted in gimbals, which in the accompanying figure are not shown. The principle employed is the maintenance of its direction in space of the axle AC of the wheel EF, when once the latter has been set spinning, and the gimbals are unlocked to allow of free motion in every direction.

The figure illustrates the method of spinning the gyroscope. The gimbals (not shown in the figure) are first locked so that the axle becomes mounted

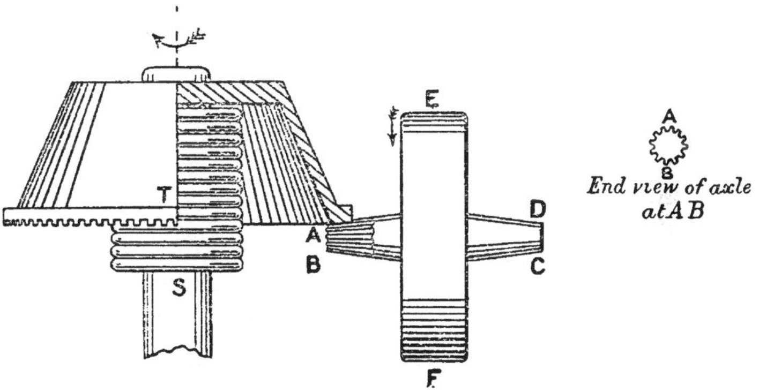

Fig. 11.

in fixed bearings. While it is in this position the teeth at the base of the segmental cone engage in the grooves of the axle at the end AB. The cone is attached to a spring S, which is automatically released as the torpedo passes through the "impulse tube." The released spring swings the cone round in the direction indicated, thus communicating rotation to the axle of the gyroscope. When the teeth at T arrive at the axle AB, *i.e.* after $\frac{3}{4}$ of a revolution of the cone, the rotation of the latter is checked almost instantaneously, by mechanism not here shown, and the gimbals are simultaneously unlocked so that the gyroscope is free to move in any direction.

The weights of the wheel (with axle) and cone are 1 lb. 12 oz. and 1 lb. 2 oz. respectively, and their radii of gyration $1\frac{1}{4}$ ins. and 2 ins.; the diameter of the cone at the base is 5 ins., and of the section AB of the axle $\frac{3}{8}$ in. If the gyroscope starts spinning at 2400 revolutions a minute, determine

(i) the energy communicated by the spring;

(ii) the average pressure between the teeth of the cone and those of the axle.

(iii) If the wheel continues spinning for 30 minutes, find the mean resisting couple due to air and friction.*

11. A small top weighing 3 oz., is so constructed that on its head is a series of small slanting flanges arranged like the sails of a windmill. The object of these is that the spinning of the top may be prolonged by blowing down on the flanges. Supposing that it is given an initial spin of 7 revolutions a sec., in the direction in which blowing *assists*, find the value of the couple due to friction at the toe and the resistance of the air when it is kept spinning uniformly under the following conditions:

There are n flanges each of area A sq. ins. inclined to the vertical at an angle of $\theta°$, the blowing pressure is P lbs. to the square in., and the distance of the centre of pressure of each flange from the axis $= l$ inches.

Supposing the initial spin is given in the wrong (*i.e.* opposite) direction, and the same blowing force applied, how soon will it be brought to rest, taking into account the resistance of air and friction, assuming their mean value to be half of that in the previous part of the question, and that the radius of gyration of the top is k ins.?

12. Fig. 12 represents a top which is spun in the following manner. The spindle projecting from the head of the top works freely in a little cap to be held in one hand, while the other pulls a short string of about 4 ins. length, one end of which is fixed in the spindle. The string is pulled out, allowed to slack and wind again on the spindle in the reverse direction. The process is repeated, and after two or three pulls the top seems to have its maximum rotation, for it starts humming. Why should the rotation be greater after, say, three pulls than after one pull?

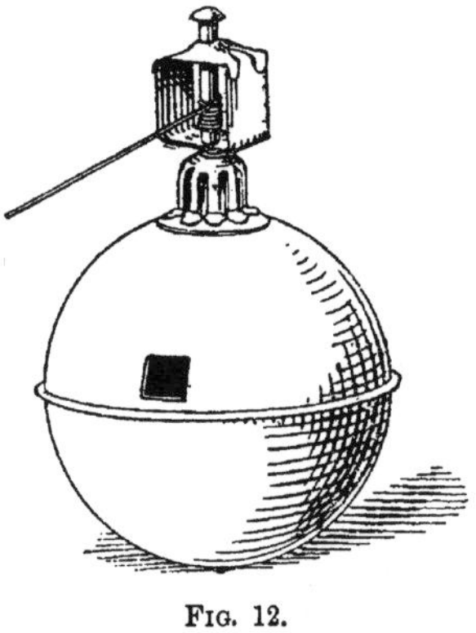

FIG. 12.

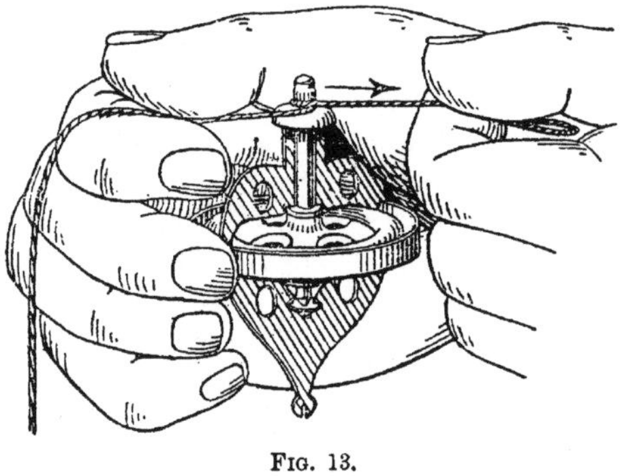

FIG. 13.

13. In another top (Fig. 13) the spindle projects from its mounting, which can be held without the top's rotation being affected; one end of the string is looped once round the spindle and this end is pulled till the whole string has been pulled over the spindle. Taking the tension in one part of the string to be $e^{2\pi\mu} \times$ the tension in the other part, where μ is the coefficient of friction and 2π the angle through which the string is wrapped, how does this method of spinning with a string compare with the more ordinary one?

---

* Further information on the steering of a torpedo is given in Chapter V. For the information there contained, and for the above approximate measurements, the author is indebted to the Admiralty authorities and to the courtesy of the Assistant Superintendent of H.M. Torpedo Factory, Woolwich Arsenal.

14. A bucket weighing 2 lbs. hangs by a rope wound round a wheel and axle, mass 8 lbs., whose radius of gyration may be taken as 15 ins. If the diameter of the axle is 6 ins., find the angular momentum of the wheel and axle after the bucket has attained a velocity of 20 ft./sec.

Express also the kinetic energy of the whole system in this case. What is the angular velocity of the wheel when 25 ft. of rope have been unwound, and how long has the unwinding taken?

It is clear that the whole system moves quicker and quicker under the action of the (constant) weight of the bucket. So when a top is set spinning by means of a string, unless the hand holding the string moves quicker and quicker the string will not remain taut. The accelerated motion of the hand does not *necessarily* mean an *increasing* pull.

15. In the above question, supposing the bucket to contain a mass of 40 lbs., find the angular momentum of the wheel and axle when the bucket has fallen 36 ft. from rest, and the kinetic energy of the whole system.

For how many seconds after this must a frictional force of $6\frac{1}{4}$ lbs. weight be applied to the rim of the axle to bring it to rest, assuming that the rope has become slack?

16. A wheel and axle of mass 6 lbs. is mounted on a table with its axis vertical, and is made to rotate by the descending of a weight of 2 lbs. attached to a string passing over a pulley at the edge of the table, the other end being wound round the axle of the wheel. The weight falls 12 ft. in 3 secs.; find the radius of gyration for the wheel and axle, if the radius of the axle is 1 in.

17. In experiments recently conducted on the *See-bar* for steadying ships at sea, a big flywheel was mounted amidships with its axle vertical. The wheel, which was in the form of a solid disc, was approximately 1 yd. in diameter, weighed (without the spindle) 1100 lbs., and was made to rotate at 1600 revolutions a minute. Neglecting the spindle, find what horse power would generate the velocity in $86\frac{2}{7}$ secs.

If steam is shut off and the flywheel comes to rest after 2750 revolutions, what is the value of the frictional couple due to the resistance of the air and the bearings?

18. Find the energy communicated to the top in example (1) by winding it up with a string a yard long and pulling

(i) with a constant force, F lbs.;

(ii) with a force which varies as the length of string (α) already used, (β) not yet used;

Comment on the similarity of answer to (α) and (β).

(iii) with a force varying as the square of the length already used;

(iv) varying as the nth power of the length yet to be used.

How would you find the angular velocity in each case?

CHAPTER II.

REPRESENTATION OF ANGULAR VELOCITY. PRECESSION.

20. If a body have a point fixed, then the angular velocity or rotation of the body has three essential properties. For·

(1) It is about a definite axis.

(2) It is in a definite direction about that axis.

(3) It is of definite magnitude.

These *three* properties can be represented geometrically, for:

(1) Through a fixed point, O say (to represent the fixed point in the body), can be drawn a straight line OA to represent the *axis of rotation*.

(2) The *direction* of rotation about this axis can be represented by considering as positive the rotation of a

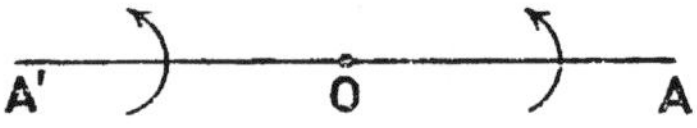

FIG. 14.

left-handed screw as one looks along the axis. For instance, in the adjoining figure the same rotation is $+\omega$ about OA or $A'A$ or $-\omega$ about OA' or AA'.

(3) Finally, we can mark off the length OA to represent on any scale we like, the *magnitude* of the rotation.

Hence it follows that angular velocity, like linear velocity, is a *vector* quantity.

21. On the physical meaning of composition of velocities. Relative and total velocity. When we say that a particle P simultaneously possesses two velocities u and v in two directions, OA, OB inclined to one another at an angle a, we do not mean that it is moving in *space* in two directions at once. That is impossible, for it moves with a definite velocity in a definite direction in space. The physical meaning of the "two simultaneous velocities" will be best understood by supposing (Fig. 15) that we have a thin wire rod, OA, on which a small bead P is moving with velocity u *relative to the end* O *of the rod* from which it started, while the *rod* moves with velocity v *in space*, keeping parallel to itself, in the direction OB. *The total velocity in space* is then along a definite straight line OD passing through the original position of O in space.

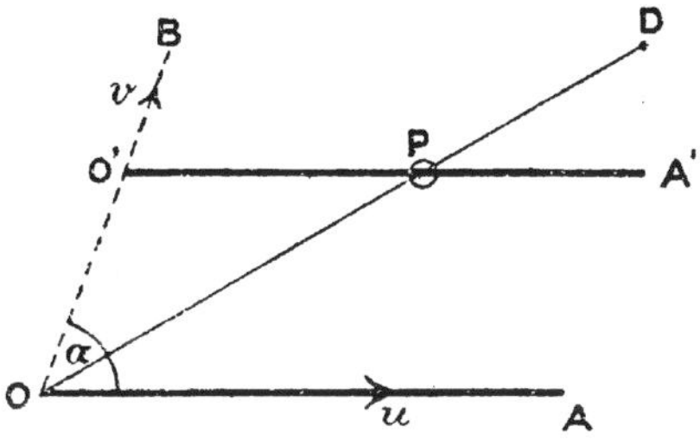

FIG. 15.

The same total velocity would have been obtained by taking the velocity v of the particle relative to a rod OB, while OB is moved parallel to itself in the direction OA with a velocity u.

22. It will be noticed that at any instant the *total velocity* of P *in the direction* $O'A'$ *in space* is equal to that relative to O' + that of O'

$$=u+v\cos a,$$

which, if $a=90°$ (namely if OA, OB are at right angles to each other)

$$=u+0.$$

Hence in this case the total velocity of P in the direction $O'A'$ in space, is the same as the *velocity relative to the rod*; so that, when we compound two velocities at right angles, each component is the *total* velocity of the particle in the direction considered. Similarly, if a particle possesses three simultaneous velocities u, v, w not in one plane, we must further suppose the whole frame AOB to move in space with velocity w keeping parallel to itself, in a direction OC not in the plane of the paper; and if the three directions OA, OB, OC, are mutually at right angles, each of the components is the total velocity of the particle in the direction considered.

23. We will now extend the above observations to the case of angular velocities or rotations.

Simultaneous Rotations. *Definition.* If a rigid body is rotating with angular velocity ω_1 about some axis OA, fixed in the body, *relative to a frame* in which that axis is fixed, while the frame rotates about some fixed axis OB with angular velocity ω_2, then the total angular velocity of the body is said to be compounded of the two simultaneous rotations ω_1, ω_2 about the axes OA, OB.

It is clear that the frame mentioned might rotate with a velocity ω_2 *relative to a second frame* which had a total angular velocity ω_3, so that the definition can be extended to any number of simultaneous rotations.

24. Parallelogram of angular velocities. *If a rigid body, of which one point O is fixed, has two simultaneous rotations ω_1, ω_2 about axes OA, OB, and if ω_1, ω_2 are represented in direction and magnitude by OA, OB, respectively, then the resulting rotation, ω, is represented in direction and magnitude by OC, the diagonal of the parallelogram of which OA, OB, form two sides.*

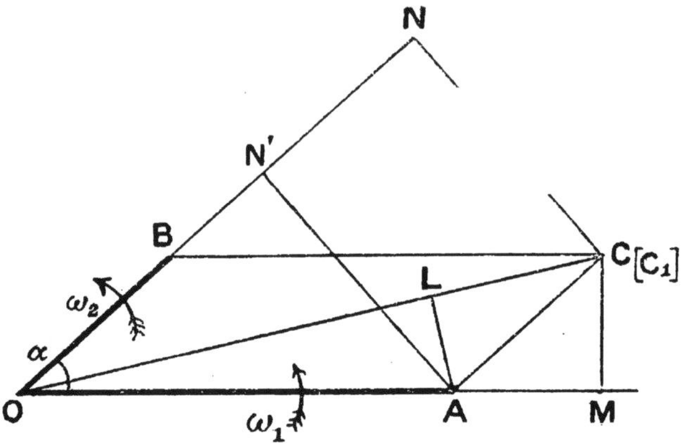

FIG. 16.

Let OA, OB (Fig. 16), represent the two rotations ω_1, ω_2 about OA, OB, respectively, the direction of rotation in each case being positive viewed from O, *i.e.* that of a left-handed screw.

Let us suppose that the angular velocity ω_1 is relative to the frame AOB, whilst ω_2 is the angular velocity of the frame itself. Complete the figure as shown.

Let C be a point in the body passing through the plane of the frame AOB and coinciding for the instant with a point C_1 in the plane of the frame.

Then the linear velocity of the point C is (away from the reader and perpendicular to the paper)

$$= \text{that relative to } C_1 + \text{that of } C_1$$
$$= \omega_1 . CM - \omega_2 . C_1N$$
$$= k . OA . CM - k . OB . C_1N,$$

where k is a constant depending on the scale chosen,

$$= k.(2\triangle AOC - 2\triangle BOC)$$
$$= \text{zero}.$$

$\therefore$ the point C is at rest (as also the point O), and therefore the body is rotating about OC, and OC represents the direction of the axis of rotation.

Again, the linear velocity of a particle at A is:

$$\begin{aligned} \omega \,.\, AL &\text{ and also} \\ &-\omega_2 AN' \\ &-k\,.\,OB\,.\,AN' \\ &-k\,.\,2\triangle\, OBC \\ &-k\,.\,OC\,.\,AL; \\ \omega &= k\,.\,OC, \end{aligned}$$

i.e. OC represents the *magnitude* of the resulting rotation ω.

Similarly, any number of simultaneous rotations may be combined by the parallelogram law. The *order* in which the rotations are combined does not affect the final result. So also any single rotation can be resolved into any number of component rotations.

It follows that angular velocities, like linear, are subject to the parallelogram law.

25. If ω_2 had been in the reverse direction originally (represented by the dotted line) we should have most easily obtained our resultant by drawing OB in the opposite direction. The resultant ω would in this case have been represented by the diagonal OC as shown in Fig. 17.

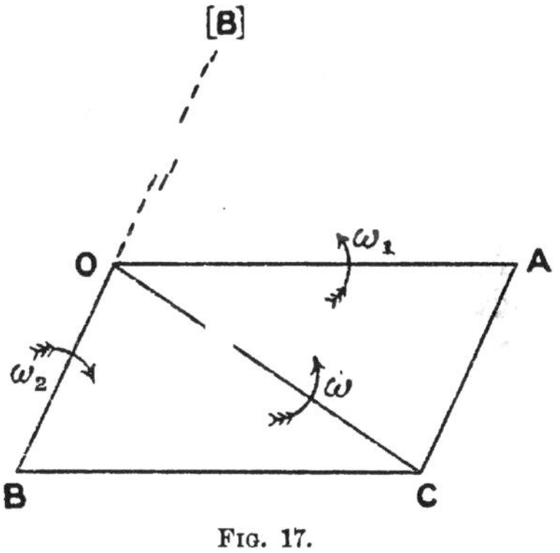

FIG. 17.

26. In practice it is sometimes more convenient to represent rotations right-handed instead of left-handed. The resultant ω of any two rotations ω_1, ω_2 will be best found by always drawing OA, OB so that ω_1, ω_2 are in the *same* direction (whether positive or negative) viewed along OA, OB; ω will then be in *this* direction viewed along OC, where OC *lies between* OA and OB.

27. The parallelogram of angular accelerations follows at once as a corollary from that of angular velocities.

28. Definition. By angular velocity about a line which is moving, we mean the total angular velocity about the fixed line in space with which the moving line happens to be coinciding at the instant under consideration.

If in Fig. 16 we consider the angular velocity ω_1 about OA to be relative to the frame, which rotates with velocity ω_2 about OB, it is clear (since angular velocities, like linear, are vector quantities) that the angular velocity *of the frame about the line* OA *of the frame* is $\omega_2 \cos \alpha$, where the angle $BOA = \alpha$.

Hence the *total* angular velocity of the body *about OA*, *i.e.* about the line fixed in space with which OA happens at the instant to be coinciding

$$=\omega_1+\omega_2 \cos \alpha$$
$$=\omega_1 \text{ if } \alpha=90°.$$

Thus, if we compound two angular velocities about axes *at right angles* to each other, each component is a total velocity—not a velocity relative to some moving plane or frame.

So too with three rotations about axes mutually at right angles.

Conversely, if we resolve a single rotation into three components mutually at right angles, each component is a total velocity, and each is independent of the other two.

The importance of the above will be seen more clearly in Chapter VI.

29. Representation of angular momentum. If we have a wheel rotating about its axle $H'OH$ with angular momentum $I\omega$, say, then using the same conventions as when representing angular velocity, we can take the line OH to represent this angular momentum. Therefore, angular momentum is a vector quantity, and hence it follows that **angular momentum**, like angular velocity, **obeys the parallelogram law.**

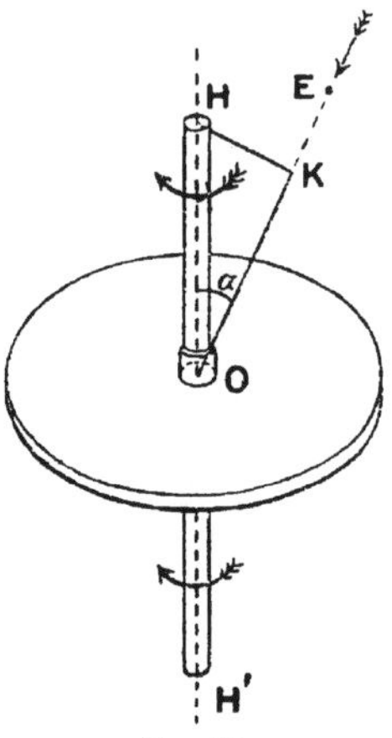

FIG. 18.

For instance: if OA (Fig. 16) be taken to represent the component angular momentum $I_1\omega_1$ of a body about OA, OB to represent the component $I_2\omega_2$ about OB, then the resultant angular momentum $I\omega$ of the body is about OC, the diagonal of the parallelogram, and is represented by OC, ω_1 and ω_2 being defined in the same way as in Art. 24.

30. It should be noted that the resultant axis of total angular momentum of a body is not necessarily the same as the resultant axis of total angular velocity: for $I_1\omega_1$, $I_2\omega_2$ are not proportional to ω_1, ω_2 respectively, unless $I_1=I_2$.

31. Referring now to Fig. 18, we see that if OE is a line making an angle α with the axle OH, then, OH being taken to represent $I\omega$ about the axle, $OH \cos \alpha$, *i.e.* OK, represents the resolute about OE, where HKO is a right angle.

We might also have taken the area, A say, of the wheel's disc to represent $I\omega$, in which case $A.\cos \alpha$ represents the resolute about the axis OE. In other words the resolute would be represented by the area of the projection of the disc as seen from a point E on OE. This method of estimating the resolute about any axis of the angular momentum of a rotating wheel is sometimes convenient.

32. Distinction between axis and axle. The axis of symmetry of a solid of revolution we shall in general call its *axle*, as distinguished from any other axis in the body about which it may be rotating or have a component rotation.

It should be noticed that the axle in this sense is a geometrical straight line—not a thick spindle relative to which, or fixed to which, the body rotates.

33. Definitions. Gyrostat. A solid of revolution which is capable of rotation about a straight rod coincident with the axis of the solid is frequently called a *gyrostat.*

Gyroscope. A gyroscope is an instrument in which a gyrostat is mounted in a frame. The most common form of a gyroscope is that in which the frame is so mounted that the gyrostat can turn in *any* direction. See Figs. XXII. sqq.

We shall, however, in general refer to the spinning wheel of the instrument as "*the gyroscope.*"

Spin of a gyroscope. By the spin of a gyroscope or top about its axle we mean the total angular velocity about the line fixed in space with which the axle happens to be coinciding at the instant in question. *It is not the angular velocity relative to the frame.*

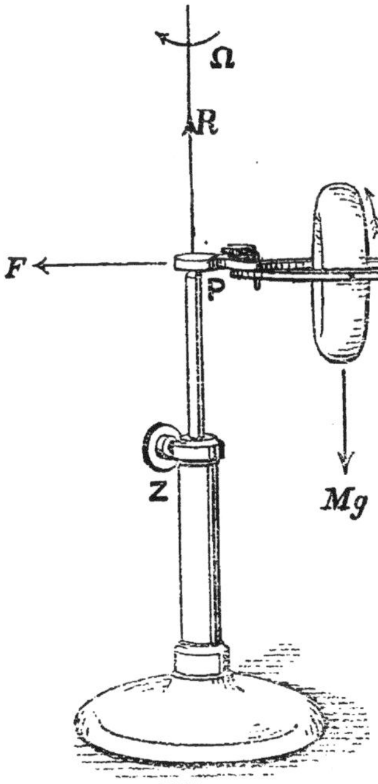

FIG. 19.

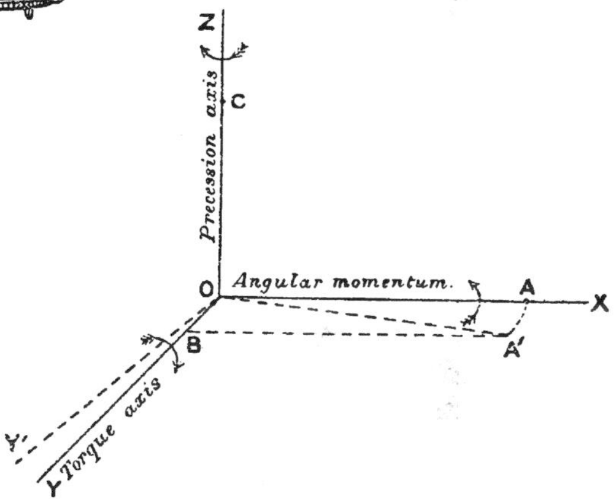

FIG. 20.

34. Motion of a gyroscope with its axle horizontal. By means of the parallelogram of angular momenta we can discuss the motion of a gyroscope which is moving, under the action of gravity only, about the fixed vertical through its point of support with its axle horizontal as in Fig. 19.

We will consider the frame to be so light that we are only concerned with the motion of the wheel itself.

Let OX, OY, OZ (Fig. 20) be three mutually perpendicular lines, OZ being vertical. Suppose the gyroscope rotating with angular velocity ω about OX, its axle, whilst the latter rotates about OZ with angular velocity Ω. The component angular momenta of the gyroscope are $I\omega$ and $I'\Omega$ where I, I' are the moments of inertia about OX and OZ. Let OA, OC represent these components in magnitude and direction.

After time δt the angular momenta are $I\omega$ and $I'\Omega$, about OA' and OC, where the angle $AOA' = \Omega\delta t$.

Then if $A'B$ be parallel to OA, OB represents the change of angular momentum in time δt, for if compounded with OA and OC it yields OA' and OC.

But $$OB = AA' = OA \times (\Omega\delta t) = I\omega\Omega\delta t;$$

$\therefore$ the change of angular momentum in time δt is about OB and

$$= I\omega\Omega\delta t;$$

$\therefore$ the rate of change of angular momentum is about OB and

$$= I\omega\Omega.$$

But this requires for its production a torque $I\omega\Omega$ about OB.

Thus the rotation of the axle about the vertical requires for its *maintenance* the application of a torque K, say, perpendicular to the vertical, and to the axle of the gyroscope, such that

$$K = I\omega\Omega.$$

35. In the gyroscope of Fig. 19, the necessary torque is supplied by the weight Mg of the system, and the equal and opposite vertical reaction at the point of support, which together have a moment Mga, where a is the distance of the centre of gravity from the point of support.

Hence, in this case, $\quad Mga = I\omega\Omega.$

Dimensions.
$$Mga = \frac{[M][L]}{[T]^2} \cdot [L].$$
$$I\omega\Omega = [M][L]^2 \frac{1}{[T]^2}.$$

It should be noticed that the centre of gravity of the system describes a horizontal circle with uniform angular velocity Ω under the action of the *horizontal* reaction at the point of support. (See Art. 43.)

36. Precession. The rotation of the axle OX of the gyroscope, by a torque about OY, round the third perpendicular axis OZ we shall call a *precessional* motion, and the axle will be said to *precess* about OZ, the axis of precession.

The equations of steady motion of a gyroscope or top with its axle inclined to the vertical are given in Chapter VI.

37. The preceding results may be stated more generally as follows, without any reference to vertical or horizontal (see Art. 85):

If a body which has angular momentum $I\omega$ about an axis OX be under the action of a torque K about a perpendicular axis OY, then the *angular momentum will be rotated* about the third perpendicular axis OZ with angular velocity Ω determined by the equation

$$K = I\omega\Omega.$$

38. Rule for direction of precession. It will be clear from the above discussion that: If the rotated angular momentum and the torque axis are drawn *in the same sense,* then the angular momentum *sets itself towards the torque axis.*

39. It should be noticed that although the gyroscope in Fig. 19 is rotating about the vertical, yet the torque does not create angular momentum about the vertical. It creates angular momentum *about its own horizontal axis,* which combining, as fast as created, with the angular momentum about the axle, causes the axle to take up successive positions in the plane XY. We have already said (Art. 34) that our investigations refer *to the maintenance of an existing motion*: otherwise our equation $K = I\omega\Omega$ involves the paradox that if $\omega = 0$, *i.e.* if the body is not spinning, then Ω, the precessional velocity, is infinite! This is, of course, not true, but it is true that a wheel spinning with a small velocity precesses with a large velocity, and *vice versa.*

A full discussion of all that occurs when precession is being started, is given in Chapter IV., where it is shown that oscillations are set up which are quickly destroyed by friction, and in many cases are scarcely visible. In instances given before Chapter IV., where, strictly speaking, the starting of precession is in part involved, the preliminary phenomena will either be assumed or neglected.

[8.] Explain what effect the couple due to the action of gravity on a Diabolo spool will have, if the string is not quite under the centre of gravity,

(i) when the spool is not spinning; (ii) when it is.

40. It is clear that the investigation of Art. 34 is equally true when K varies and consequently Ω as well.

For let θ be the angle which the axle makes at any instant with its initial position.

Then $$\delta\theta = AOA' \text{ and } AA' = I\omega\delta\theta.$$

Hence the increase of angular momentum in time δt is about OB and is equal to $I\omega\delta\theta$: namely, the rate of change of angular momentum is equal to

$$I\omega \cdot \frac{d\theta}{dt} = K.$$

It follows that, if ω is constant,

$$\theta=\frac{1}{I\omega}\int K dt,$$

or the precessional angle swept out during any time is a measure of the time-integral of the applied couple.

41. Analogy in linear motion. The application of a torque to maintain the steady rotation of the axle of a spinning body about a perpendicular line has its analogy in linear motion, where a force is needed to maintain the steady rotation of the line of motion of a particle moving uniformly in a circle.

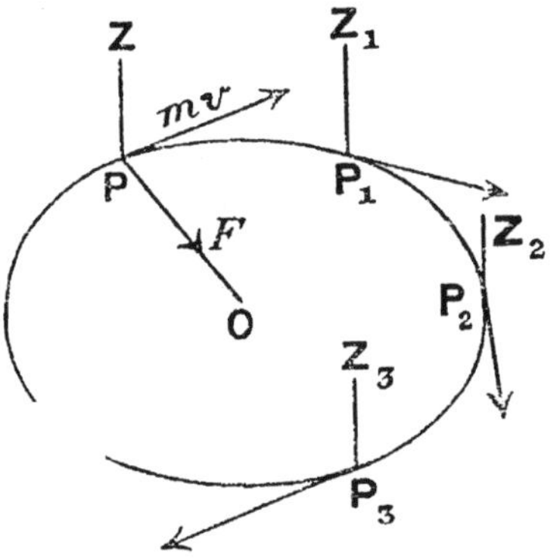

FIG. 21.

For let a particle P, of mass m, move with constant velocity v round a horizontal circle whose centre is O, towards which is a constant force F.

Let P_1Z_1, P_2Z_2, P_3Z_3 (Fig. 21) be successive positions of the vertical through successive positions of P. Then the effect of applying the force F perpendicular to the line of the momentum mv of the particle, is to *turn* the line of momentum *at a constant rate* about an axis perpendicular to F and itself.

In this case, if the rate at which the line of momentum is being turned be Ω, then Ω is equal to the angular velocity of the particle about O, *i.e.* $\Omega=\frac{v}{r}$, and we have

$$F=\frac{mv^2}{r}$$

$$=mv\,.\,\Omega.$$

42. Diabolo. The rule given above (Art. 38) for the direction of precession will at once explain the reason for the instructions (variously worded) which are given to beginners learning to spin a Diabolo spool.

Let us suppose that the Diabolo is being spun right-handed as in the illustration (Fig. 22), when the end farther away from the performer begins to dip down, and the nearer end to rise. To get the spool again horizontal it is necessary for the performer to pull back his right hand and (in imagination) press down the rising end of the spool with what may be called the "working" end of the string. It will thus be seen that a torque is created about an axis perpendicular to the axle of the spool, and since the axle tends to set itself towards the torque-axis (both being drawn in the same sense), the lower end of the spool rises, as illustrated in Fig. 23. The same principle holds if the end next the performer is dipping; also if the spool is being spun left-handed; and a general rule

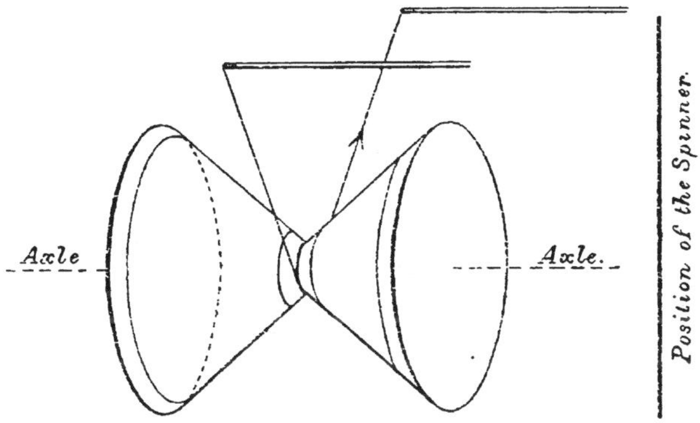

FIG. 22.

could be formulated to the exclusion of all ideas of left- and right-handed by saying that the working end of the string must be moved so as (in imagination) *to press down the rising end.*

Let us now suppose that it is required to make the Diabolo turn round to the right of the performer when it is being spun as in Fig. 22. It is clear that a torque applied by the strings about a vertical axis to the right, as viewed from above, will result in the near end *dipping.* A torque must be applied about a *horizontal* axis, and taking the rotation of Fig. 22, this torque can be applied in the right "sense" by the performer dragging the spool slightly towards him, or (without moving the spool) by changing the point of support so as to make it slightly nearer to himself. If the spool is dragged from the under side it is advisable to extend the two strings at a large angle, as otherwise the dragging action will tend to depress the end nearest to the performer and thus produce the opposite effect to that aimed at.

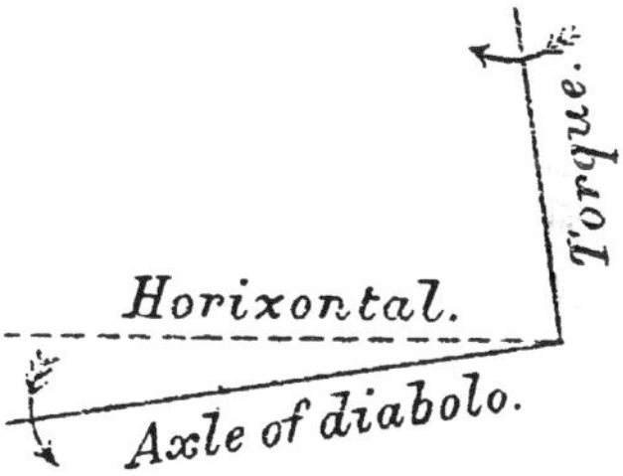

FIG. 23.

[9.] If the Diabolo is being spun left-handed, what is the effect caused by the performer

(i) drawing in his right hand?
(ii) drawing in both hands?
(iii) drawing in his left hand when the end near him begins to dip?

A discussion is given in Chapter IX., on a Diabolo which will not spin.

[10.] In a motor-car the fly-wheels are mounted so that their *axles* are parallel to the direction and motion of the car. Supposing the fly-wheels to revolve right-handed when viewed from the back of the car, what is their effect on the car as it turns a corner (*a*) to the right, (*b*) to the left?

[11.] (*a*) When a paddle-boat is steaming ahead, what is the gyroscopic effect of paddle-wheels when a wave strikes the ship and gives it a list to starboard?

(*b*) What is the effect of steering to starboard?

EXAMPLES.

1. In the equation $K=I\omega\Omega$, if ω is 20 radians a second, $\Omega=5$ radians a second, $I=50$ lb.-ft.^{2}, what is the value of K in ft.-lbs.?

2. The moment of inertia of a heavy wheel about its axle is 15,000 lb.-ft.^{2}. If it is making 3,000 revolutions a minute, what will be the velocity of precession when a couple of 12,000 ft.-pdls. is applied to the axle of the wheel?

3. Find the length of the axle of the wheel in Fig. XXII., supposing its moment of inertia to be 3,000 gram.-cm.^{2}, and that, when spinning at the rate of 50 turns a second, a weight of 60 gr. suspended from X causes it to precess at 3 turns a minute.

4. Find the rate of spin of a gyroscope, taking the length of its axle as 12 cm., the moment of inertia 3,500 gr.-cm.^{2}, the rate of precession 3 turns in 70 secs., and the suspended weight as 75 grms.

5. A heavy wheel, mass 50 lbs., in the form of a disc, is rotating about its axle with velocity 16 radians a second, and precessing at the rate of 4 radians a second under the influence of a couple of 100 ft.-lbs. Find the radius of the wheel.

43. Independence of translation and rotation. It is proved in works on rigid dynamics that any motion of a rigid body can be completely represented by a translational motion of its centre of mass G combined with a rotational motion about some axis through G; and further, that these two motions are independent of each other and produced independently by the acting forces.

Hence, in considering the motion of any rigid body, whether some point in it is fixed or not, we may discuss the rotation of the body as if G were fixed, and the translation of the body as if it were a single particle, the entire mass being concentrated at G.

The truth of this can be illustrated by taking an ordinary walking stick and marking the position of its centre of mass with a piece of chalk. If the stick is then suspended loosely from the fingers of one hand while a blow is given to the stick, it will be found that, whenever the blow is struck so as to pass through the centre of mass, the stick flies out of the hand with a motion of translation only, that is to say, it has no rotation. If, on the other hand, the blow is struck through *any other point* than the centre of mass, then it will be observed that not only does the centre of mass have a translational motion, but the stick also rotates about this (moving) centre of mass.

A familiar illustration of this independence of translation and rotation can be given by tilting back a chair and then letting it fall forward again. The chair not only swings forward to its original position, but also *moves forward along the carpet*: for, as it swings forward, its centre of gravity has a forward translational motion, and, before the chair can come to rest, this motion has to be stopped (by the friction on the carpet), independently of the rotation. The rotation is, of course, stopped by the moment of the vertical reaction of the floor on the two front legs about the line joining the points of contact of the two hind legs with the floor.

44. Further illustrations of precession.

Hoop. If a wheel, or an ordinary hoop, is rolling along the ground it has an angular momentum about its central perpendicular axis or axle. If the hoop begins to move in the slightest degree out of the vertical plane, the external forces acting on it (*i.e.* the reaction of the ground and the weight of the hoop) tend to tilt over the axle, with the result that it "precesses" about an axis very nearly vertical, and its path becomes a curve instead of a straight line.

Engine wheels. The tilting moment may be considerable without the wheel leaning over at a large angle to the vertical, as in the case of wheels of a railway engine where the reaction of the rails on the wheels is very large. Let us suppose the engine turning to the right, looking in the direction in which the train is going. To make each wheel precess to the right, a couple must be applied about a horizontal axis (through its centre of gravity, say) parallel to the direction of motion of the train; and this couple must, as we have seen, be in a direction tending to lift up the left-hand end of the axle and to depress the right-hand end. Therefore, *owing to the rotation of the wheels* (apart from any motion of translation), it is necessary to raise the left rail slightly in rounding the curve. If this were not raised the whole turning of the train would be effected by the *lateral* pressure on the left-hand flanges, which (in addition to considerable wear to rails and wheels) would tend to make the wheels precess *to the left*, and would increase the risk of their "jumping the metals."

Gyroscopic action in a bicycle. The wheels of an ordinary bicycle have rims so light compared with the rest of the machine that their angular momentum does not play a preponderating part in the motion of the whole machine. At times, however, it is quite an appreciable quantity, as can be easily realised by taking the front wheel out of its bearings, holding the axle in the hands and attempting to turn the axle after a considerable spin has been given to the wheel.

There is a certain amount of gyroscopic action brought into play when the rider, finding himself falling over, say to the right, gives the handle bar a sharp turn to the right. In this case we shall see, if we consider the directions of the rotations in question, that this twisting of the handle bar causes the front wheel to precess about a horizontal axis into a more vertical position, *i.e.* to right itself. Simultaneously with the application of the couple to the front wheel by the man there is a counter couple applied to the man, and therefore to the back wheel, which causes it to turn at a *greater* inclination to the vertical, and so to increase the curvature of the path of the bicycle.

The preponderating cause, however, of the bicycle righting itself is to be found in the *linear* momentum of the whole machine and of the rider, for by far the greater part of the motion is *not rotational.*

When the front wheel is turned at an angle to the back wheel, the linear momentum of the rear part being checked by the front part brings a force to bear on the latter, one component of which drives it along its path, while the other, perpendicular to the wheel, raises it to a more vertical position, and consequently the whole cycle proceeds in a more vertical position.

It is clear that if the brake be applied to the front wheel and not to the back, the bicycle will tend to skid more than if the linear momentum of the rear part is checked while the front wheel is free to move along its path.

Motor cycle. In a motor bicycle there is a heavy fly-wheel rotating rapidly in the plane of the road wheels, so that as the cycle rounds the curve the fly-wheel is constrained to precess, and therefore, on account of *this fly-wheel alone,* a couple $I\omega\Omega$ must be applied to cause this precession, apart from the other forces required to turn the rest of the machine. In other words, the rider must lean over more than if the fly-wheel were not there, or were not rotating. In practice this wheel rotates in the same direction as the road wheels.

This necessity for leaning over an extra amount accounts in part for motor bicycles being more liable to skid when going round a corner than ordinary bicycles.

45. Gyroscopic resistance. We have seen that in order to turn the axle of a rotating wheel about a perpendicular axis, it is necessary to apply a couple K, about a *third* axis, perpendicular to the other two, whose measure (employing the above notation) is $I\omega\Omega$. Thus the *resistance* which the axle of a rotating wheel *offers to being turned* about the axis of the applied couple K is measured by $I\omega\Omega$, and this expression is frequently called the *gyroscopic couple*, or *gyroscopic resistance.* Hence, the only difference in discussing the motion of a rotating body, as compared with that of the same body not rotating, is that we must include the gyroscopic couple among the resisting forces.

As a further illustration of the gyroscopic couple, it will be noticed that if we try to turn the axle of a wheel which is *not* rotating we meet with no resistance beyond that of the inertia of the body, *i.e.* $\omega=0$, and the gyroscopic couple $I\omega\Omega$ becomes zero. So in the Introductory Chapter, when the gyroscope (Fig. XXII.) is prevented by the clamp at Z from turning about the vertical, *i.e.* when $\Omega=0$, the gyroscopic couple is again zero and the body offers no resistance to being turned about YY', except that of inertia. Hence we see that *a rotating wheel will offer no gyroscopic resistance unless it is free to precess.*

Again, when we apply the equation $Mga=I\omega\Omega$, in Art. 35, to obtain the equation of steady motion of the gyroscope, we are expressing the fact that the gyroscopic couple $I\omega\Omega$ is balancing the couple Mga due to gravity.

[12.] What corresponds to the gyroscopic couple in uniform circular motion?

If we neglect the inclination to the vertical of a wheel as it rolls round a curved path, the expression for the gyroscopic couple simplifies. For if v is the rate at which the wheel is travelling, r its radius, R the radius of the curve,

$$\omega=\frac{v}{r},\quad \Omega-\frac{v}{R},$$

$$\text{and}\quad I\omega\Omega=\frac{Mk^2v^2}{R\,.\,r},$$

k being the radius of gyration.

The investigation when the inclination of the wheel to the vertical is taken into account, will be found in Chapter VI.

FURTHER EXAMPLES.

1. A metal disc of radius 1 ft. and mass 2 lbs., is made to roll uniformly round a curve of radius 12 ft. The gyroscopic couple is 5 ft.-lbs. Neglecting the inclination of the disc to the vertical, find the rate of rolling.

2. An engine on six wheels, each of radius r, is rounding a curve, radius R. If V is the speed of the train, M the mass of each wheel, k the radius of gyration, and $2h$ the width of the gauge, show that the gyroscopic couple due to the wheels is

$$\frac{6Mk^2RV^2}{r(R^2-h^2)}.$$

Owing to their rotation, which wheels tend to come off the rails?

3. If the wheel of the gyroscope in Fig. XXV. is a solid disc spinning with velocity ω, and precessing with velocity Ω, find its radius if $2a=$ the length of the axle.

4. The propeller shaft of a Torpedo-Boat Destroyer is a cylinder of radius a ft., and weight M lbs. When the shaft is revolving with an angular velocity ω the speed of the ship is V ft./sec., and the radius of the turning circle is R ft. Find the torque exerted by the bearings of the shaft.

5. In the preceding question, taking $a=4$ ins., $M=12$ tons, $V=35$ knots, $\omega=750$ revolutions per minute, $R=605$ yds., estimate the torque in ft.-lbs. These values are taken from H.M. T.B.D. "Tartar."

6. A motor cycle has a heavy fly-wheel, in the form of a disc of radius 4 ins. and weight 32 lbs., which is making 1,600 revolutions a minute. The machine and rider weigh 320 lbs. If the rider rounds a curve of 25 ft. radius when travelling at 20 mi./hr., find approximately the horizontal distance that he must move the centre of gravity of himself and his machine *owing to the rotation of the fly-wheel.*

NOTE. If the fly-wheel were not spinning he would have to lean over a certain amount in order to turn the corner; but when the fly-wheel is spinning, he must increase the arm of the gravity couple by a small amount, x, in order that the fly-wheel may precess. It may be assumed that the gyroscopic couple is the same as if the fly-wheel were in a vertical plane.

7. In the previous question, just before the rider rounds the corner he leans over to the right with an angular velocity of a radian a second. (*a*) How is the torque supplied which causes the fly-wheel to have this angular velocity? (*b*) About what axis was it, and in what direction? (*c*) What is the effect on the back wheel of the reaction to this torque? (*d*) Find its magnitude.

CHAPTER III.

DISCUSSION OF THE PHENOMENA DESCRIBED IN THE INTRODUCTORY CHAPTER.

46. If we turn to the Introductory Chapter we shall now see the reasons for the paradoxical behaviour of the tops there described. To take the first instance mentioned, if a top is placed on its toe while spinning, the torque, which would cause it to fall over if it were not spinning, fails in this case to turn over the axle but rotates it, and so the top precesses. Also, if we consider the rule given in Art. 38 for the direction of precession, we see why an ordinary top precesses in the direction of its spin, when viewed, say, from above (Fig. I.), while that in Fig. III. (*a*) precesses in the reverse direction· for in the latter case the centre of gravity is *below* the point of support and the torque acts in the reverse direction.

47. Let us now consider what causes a top that has been spun at an inclination to the vertical to rise to a more vertical (though not necessarily quite vertical) position before settling down to a steady motion. Suppose we take an ordinary top and spin it on a rough board. When it has settled down to a motion in which its axle is practically at a constant angle to the vertical, let us drag the board round in the same direction as the precession. It will be found that the centre of gravity of the top rises, and some of us may remember that the method of resuscitating a dying top is to get it on to one's hand and drag the hand round in the necessary direction. Similarly, if we draw the hand round in the opposite direction the top falls; so that we can sum up our experiment in the two following statements:

(i) hurry the precession, the top rises;

(ii) retard the precession, the top falls.

Hence, when a top is spun at an inclination to the vertical and rises, there must be at first something that is hurrying the precession, and eventually (since the top afterwards falls) something that retards it.

Now the only forces acting on the top (besides the resistance of the air) are:

(i) its weight;

(ii) the reaction at the point of contact of the toe with the table.

The weight acting vertically downwards cannot hurry or retard the precession.

In Fig. 24 let K and A be respectively the centre and radius of curvature of P's path. The reaction at the toe can be resolved into two components, one S, *perpendicular to* the plane on which the top is spinning, due to the *stiffness* of the plane:

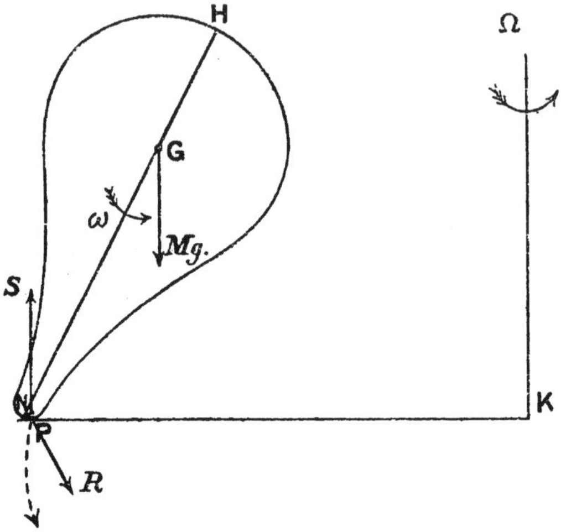

FIG. 24.

and another, R, *in* the plane due to its *roughness*. The component of R in the direction PK provides the necessary normal acceleration, while the component along the tangent to the path is either hurrying or retarding the precession.

If it is hurrying the precession we see by the rule given in Art. 38, that the axle of the top rises * to a more vertical position, and *vice versa*. Let us consider the cases in detail.

Let ω be the velocity of spin, and Ω the angular velocity of P about K, and let $PN=a$. The direction of the friction at P depends on the direction of motion of P, for it will act contrary to the direction of the latter. At the start there must be skidding if the top is dropped on to the plane.

Now the actual velocity of P is that relative to N+that of N; or, the actual velocity of P (measured *into* the paper)

$$=a\omega-A\Omega.$$

* The rise or fall of a top is more fully discussed in Chapter IV. from considerations of conservation of angular momentum and conservation of energy.

But at the start ω is very big, though it gradually diminishes under the action of the friction, and also the resistance of the air.

Hence, at the start we have

$$a\omega > A\Omega, \quad \text{............................(i)}$$

i.e. the top is skidding and P is moving *into* the paper; friction accordingly acts in a direction which hurries the precession, and the top therefore rises.

In the meantime ω is diminishing (and also a as the top gets more vertical), and after a time we have

$$a\omega = A\Omega, \quad \text{............................(ii)}$$

i.e. P has no motion relative to the table, and steady rolling motion becomes possible.

And, later, we have $a\omega < A\Omega$, ...(iii)

i.e. friction is retarding* the precession and the top begins to fall.

Were it not for the resistance of the air, the state of steady motion, when once established, would, theoretically, continue indefinitely: for as soon as $a\omega$ becomes $= A\Omega$ rolling commences, which can only be disturbed by air resistance. Immediately $a\omega$ becomes $< A\Omega$ friction acts so as to *increase* ω, but *never beyond the point necessary to maintain steady motion*, since it is a self-adjusting force. Hence, the position of steady motion is a stable one, for friction produces an oscillating tendency to return to it; but eventually, the resistance of the air, tending to destroy this state, overcomes the tendency of friction to restore it.

[13.] Even *in vacuo* the top would not really spin for ever. What is the explanation of this?

[14.] In Fig. XVII. why must the gyroscope be placed on a smooth surface?

[15.] The rougher the table the more quickly the top will rise when first spun. Why?

[16.] Given a rough table, why does a top with a blunt peg rise more quickly than one with a fine peg?

[17.] Explain the apparent contradiction of the tight and loose screw in the top of Figs. IV. (a) and IV. (b).

[18.] Supposing our object in spinning a top is to make it "go *for as long as possible*," how would

(i) a blunt peg, (ii) a fine peg,

contribute to length of time?

[19.] How would the centre of gravity behave if the table were perfectly smooth?

---

*It will be shown in Chapter IV. that, as the top dies, though friction is *resisting* precession, yet owing to other causes, Ω increases. Amongst these is the fact that the gravity couple acts on a longer arm.

[20.] Should we get precession? If so, how would the axle move when the motion is steady?

[21.] What is the condition that a top should *fall* to its steady position? What conduces to this condition?

[22.] Discuss the rising of that form of top where the "body" to which the spin is given revolves freely on a spindle carrying the toe; namely, where the toe and spinning body are not rigidly connected. Does it arrive at steady motion more quickly or more slowly than the other kind?

48. The discussion given in Art. 47 explains the behaviour of the whip-top and loaded sphere described in the Introductory Chapter, as also that of the acorns and hard-boiled eggs: for the effect of friction at the point of contact is to hurry the precession, and so raise the centre of gravity, as in the case of the top. The phenomenon exhibited by the top of Fig. VI., and by the egg of Fig. IX., is caused by a tendency of the liquid to spin about its *shortest* axis, which overcomes the effect of friction tending to raise the centre of gravity and so bring the top on to its longest axis. To understand this thoroughly it will be necessary to consider under what conditions a body will spin about its shortest axis in preference to any other.

49. Tendency to spin about the least axis. If a rigid body, with three perpendicular axes of symmetry, is free to turn in any direction about its centre of gravity G, and rotation be continuously given to the body about a fixed direction through G, for example the vertical, *but no other forces act on it*, the body will set itself so that its *least* axis through G becomes vertical, and it will spin stably in this position.

For every particle composing the body tends (owing to what is frequently called centrifugal force) to separate itself as far as possible from the vertical axis, and so increase the moment of inertia about that axis. Hence the body will only spin stably when the moment of inertia is as great as possible; namely when the *least axis* of the body through G is in a vertical position.

This can be easily illustrated by taking a stone or any other rigid body attached to a string and, starting with the string vertical, whirling the body round and round oneself as a vertical axis. The body will rise higher and higher until the string is horizontal; but having arrived at this position it will continue to maintain it, whatever additional spin may be communicated about the vertical axis.

Similarly, it will be seen that a liquid or viscous body which is being made to rotate about a vertical axis will tend to *change its shape*, in such a way that its vertical axis becomes smaller; and if there are no external forces tending to rotate the body, but rotation has already been communicated about some axis, this axis will tend to become smaller and smaller, as is the case with the Polar axis of the Earth.

50. We shall now be able to see the reason for the behaviour of the eggs and the tops which were full of liquid.

We know (Art. 47) that the shell, even though originally spun about its shortest axis, would, if empty, rise and spin about its longest axis, owing to the friction at the point of contact with the table. But between the liquid and the shell there is comparatively little friction, so that the liquid, once spun about its shortest axis, continues to spin about that axis; and being of much greater mass than the shell it overcomes the latter's tendency to rise on to its longest axis, with the result that the whole body, shell and liquid, spin about the shortest axis.

It is clear that the final behaviour of the egg, or top, must depend, amongst other things, on the relative masses of the liquid and shell, and this accounts for the top of Fig. VIII. spinning about its longest axis although full of water.

51. The gyroscopic top (Figs. XV. *a*, XV. *b*) is another, and perhaps more remarkable, instance of the precession of the axle.

The forces acting on the top are (Fig. 25):

(i) its weight;

(ii) the normal reaction *S* at the point of contact of the top with the spiral coil;

(iii) the tangential reaction *F* at that point;

(iv) the reaction at the point of support *O*.

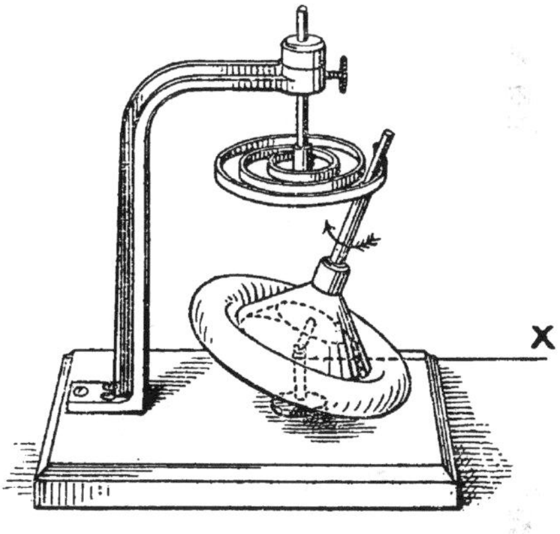

FIG. 25.

In most models of this top, the centre of gravity is made to coincide with the point of support, though in some it can be adjusted so as to be either above or below as required.

For the sake of simplicity we will consider that it is coincident with the point of support—though the following explanation would only require a little modification if this were not so.

Let us regard the axle of the top as in the plane of the paper, and let the spindle be rolling (left-handed as viewed from *O*, the point of support) along the inside of the coil, into the paper, and be approaching the end of the coil. The motion will be most easily explained by considering first the effect of the tangential reaction *F* at the point of contact *P*, and then the additional effect of the normal reaction *S*.

(1) *Tangential reaction.* It is evident that the friction acts into the plane of the paper and perpendicularly to it. This

friction, together with the equal reaction at O, forms a couple about the horizontal OX, which lies in the vertical plane through the spindle, and consequently the axle of the top tends to set itself *towards* OX, *i.e.* towards the coil, downwards, but is prevented from doing so by the normal reaction S of the coil.*

The spindle, in consequence, *pressing hard against the coil*, and thus increasing the friction, rolls or skids along until it reaches the end, when, no longer meeting with any resistance from the coil, it rushes rapidly round the corner. After this it *presses up* against the coil, since friction now acts out of the paper towards the reader, and the motion is continued on the same principle as before.

(2) *Normal reaction.* The effect of the reaction S is to accelerate the motion of the spindle *along the coil.* For, if we again take the position of Fig. 25, it is evident that S creates a torque about an axis drawn from O *into* the paper, towards which axis the spindle sets itself by the laws of precession.

It should be noticed that if the centre of gravity of the top is at the point of support, the top, when at rest, will balance with its axle at any inclination to the vertical. In this case, even though it is touching the coil, there is no reaction at the point of contact; but the moment the top is spun, any contact with the coil involves friction, and therefore a precessional tendency against the coil.

52. The motion of the toy gyroscope in Figs. XVI.–XXI. is due to the action of the external torque which causes the axle to precess, but there still remain one or two points worthy of remark in connection with the scientific gyroscope of Fig. XXII. It was mentioned that if, while the gyroscope is spinning with the weight attached at X, the screw at Z be tightened, then the gyroscope at once turns about $Y'Y$. It is interesting to note that the direct cause of this turning is *not the moment of the weight about* $Y'Y$. The moment of this weight tends to set up precession about ZZ', but since the screw at Z is tight, and consequently the frame containing $Y'Y$ is fixed, the pivots at Y and Y' meet with resistance forming a left-handed couple about $Z'Z$, and, as in the case of the gyroscopic top, it is *this* resisting couple about $Z'Z$ which turns the gyroscope about $Y'Y$ in accordance with the law of precession. Similarly, referring to Fig. XXIII., it is this resisting couple about a vertical axis which causes the gyroscope to set itself with its axis vertical, and this couple reversed which brings about the somersault.

53. An explanation of the oscillations mentioned in the Introductory Chapter on page 10 will be found in Chapter IV.

* This normal reaction is of very considerable magnitude, for it is found that the coil, unless strongly constructed, becomes bent out of shape.

54. We will now discuss why the "drifting" in still air of a rifle bullet is due to gyroscopic action, while the swerving of a "sliced" golf ball is not. The essential difference is that in the case of the rifle bullet the axis of spin is initially a tangent to the path described, while in the case of the golf ball it is a normal. Let us take the latter case first.

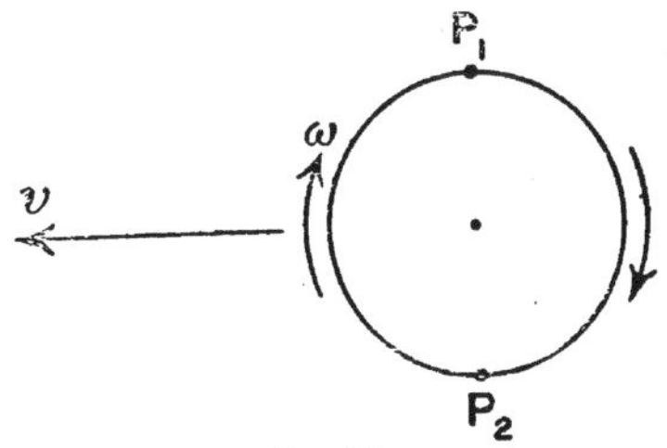

FIG. 26.

Suppose that the ball is travelling at velocity v, with a spin ω due to "slicing" in the direction indicated in the figure, which represents a horizontal section of the ball. Let its radius be a. Consider the respective velocities of two points P_1, P_2 at opposite ends of a diameter which is at right angles to the path of the centre of the ball.

$$\text{The actual velocity of } P_1 = v - a\omega.$$
$$\text{,, \quad ,, \quad } P_2 = v + a\omega.$$

But the pressure of the air on the surface of a moving body increases with the velocity. Hence the pressure of the air on P_2 is greater than on P_1, and consequently the ball will swerve to the right, looking in the direction of motion.*

In the case of the rifle bullet, however, the swerving is in the first instance due to gyroscopic action. The direction of the bullet's spin (due to the rifling of the barrel) is left-handed, viewed from the gun (Fig. 27). The excess of wind pressure

FIG. 27.

on the under side of the nose-end of the bullet *tends* to tilt the point upwards, but results instead, owing to gyroscopic action, in the bullet working over slightly to the left and becoming slightly "broadside on" to the direction of its path. This is not shown in Fig. 27. The consequent increased air

---

* It is for this reason that the face of a "driver" is slightly "laid back"; for, since the ball is "teed up" before being driven, and the club strikes it horizontally, the inclination of the face to the vertical should give a *slight under-spin* to the ball, which we see from the preceding, causes it to *rise against gravity*. Otherwise, with the same initial velocity, it would be impossible for the ball to travel as far as it does.

It is said that the late Professor Tait, having been given initial data, declared it to be impossible to drive a golf ball beyond a certain distance, which he specified. When his son, the late F. G. Tait, exceeded this distance by a considerable number of yards, the Professor examined the problem more thoroughly and arrived at the above explanation.

pressure on the right-hand side causes the bullet to "drift" to the left.

In the case of a boomerang the varying air pressure is due to both of the above causes and to others of a more complicated nature.

55. Celts. Referring to the Celts described in the Introductory Chapter, the explanation of their behaviour is to be found in the direction of the normal reaction at the point of contact with the smooth horizontal plane.

In Fig. XI. (Plate I.)* the surface near the point of contact is approximately spherical, so that when the stone is tilted in any direction the normal reaction passes through the line GC and so has no moment about GC. Hence the stone will have no tendency to turn about GC.

But in Figs. XII., XIII., this is not so, and phenomena (a) and (b) are explained in the following way:

Phenomena (a). Each base is in these two cases roughly ellipsoidal near the point of contact, but the ellipsoidal surface is "slewed round" slightly, relatively to the main body of the stone.

Accordingly, when the stone is slightly displaced from its position of equilibrium, the vertical reaction, which is normal to the ellipsoid at the point of contact, does not pass through GC.

Hence it has a moment about GC tending to turn the stone about GC.

In Fig. XII. the vertical taps at A and at B result in moments of opposite sign about GC; in Fig. XIII. they result in moments of the same sign, *i.e.* from A to B.

Phenomena (b). When the stone is spun about the vertical GC, since the conditions are not absolutely perfect, small oscillations will be set up, and the normal reactions already described will be called into play.

These may or may not tend to stop the rotation, and so they may either increase or diminish the oscillations.

The particular effect of the normal reaction will of course depend on the special shape of each stone.

In any of the above cases, the motion, whether reversed or not, will only cease when the original energy of spin has been destroyed by friction or communicated by oscillation to the horizontal surface on which the spinning takes place.

A complete analytical investigation of the above phenomena is given by Mr. G. T. Walker in *The Quarterly Journal*, No. 110, 1896.

---

* For Plates I.-III. see end of book.

CHAPTER IV.

THE STARTING OF PRECESSION. OSCILLATIONS OF THE GYROSCOPE.

56. Conservation of angular momentum. We know that when two particles collide, the total momentum of these two particles, measured in any direction, is the same after impact as before; for the impulsive reaction between the two produces as much momentum in one particle in one direction as it does in the other particle in the opposite direction. Hence, the moment of the momentum (*i.e.* the angular momentum) about any fixed line must be the same after impact as before. It follows therefore that, in any system of particles whatever, whether rigidly connected together or not, if internal forces only, and no external forces, act on the system, then, however the configuration of the particles may change, the total angular momentum of the system (*i.e.* $\Sigma mr^2\omega$) about any fixed line in space remains constant.

ILLUSTRATIONS.

(1) If a mouse is put in a cylindrical cage capable of revolving freely about a horizontal axle, then, in whatever way the mouse moves, the total angular momentum, about the axle, of the mouse and cage together, remains constant. It is clear to an observer that if the mouse climbs directly up the curved surface of the cage, the cage begins to revolve in the opposite direction, though not necessarily with the same angular velocity as the mouse; for though $I_1\omega_1$, the angular momentum produced in the mouse, is equal to $I_2\omega_2$, the angular momentum produced in the cage, I_1 is not necessarily equal to I_2.

(2) If a horizontal disc were constructed which could turn perfectly freely about a vertical axle, a juggler spinning on the disc could diminish or increase his rate of rotation by extending or dropping his arms, and thereby increasing or diminishing his moment of inertia. A similar action enables a cat to right itself while falling in mid air, legs uppermost, and land on its feet. (See instance 5.)

(3) If a skater is describing a circle about the orange as centre, he can increase his angular velocity about it by leaning over directly towards it, although there is no tangential force tending to increase this velocity. For his moment of inertia about the orange becomes smaller, but his angular momentum about the vertical through the orange remains unchanged.*

If he describes the circle with angular velocity, ω say, he has also an angular velocity ω about the vertical axis through the point of contact of his skate with the ice; for he makes a revolution about this axis in the same time that his centre of gravity makes a revolution about the orange. The pressure exerted by the ice has no moment about the vertical axis through the point of contact. If, while he is describing the circle, he extends his arms, he increases his moment of inertia and therefore diminishes his angular velocity ω about the axis in question; and this diminution of ω is shown on the ice by a "flattening" of the curve described—a necessary result, since ω is also the angular velocity of his centre of gravity round the orange, which, if diminished, involves a larger radius for the curve.†

(4) In the case of the earth and the moon we have another instance of the conservation of angular momentum. The action of the tides on the earth tends to diminish the kinetic energy of the latter, and therefore its rate of spin; but the *angular momentum* of the whole system, earth and moon, remains unchanged. Hence, since the angular momentum of the earth itself is diminished by the tides, it follows that the moon must recede from the earth so as to keep the angular momentum of the whole system constant.

(5) **Cat righting itself in mid-air.** Suppose the cat to start falling vertically with no rotation, back downwards and legs extended at full length perpendicular to the body as depicted in Plate II. Then during its fall the animal turns itself through 180° about some axis, roughly a line running lengthways through the centre of gravity of its body; but since it starts with no rotation, it can at no time during its fall have any angular momentum about this line, which we will call its axis.

(An examination of the photographs given will exclude the idea, sometimes suggested, that the cat *gives itself rotation* by using the hands for a fulcrum just as it is let go. The first few images show no tendency to turn either to one side or the other.)

---

* In the same way, when water is rotated and then let out of a hole at the bottom of a basin, as the distance r of any particle from the hole diminishes, since the angular momentum remains constant, ω increases and the water goes round with increasing speed.

The same is true of a stone tied to one end of a string and swung round horizontally while the other end of the string winds up round a fine stick or peg.

† It should be remarked that the above reasoning, although in agreement with experiment, is nevertheless not strictly accurate, since we are considering angular momentum *about a moving line.*

Let us regard the cat as made up of a fore part and a hind part, whose moments of inertia I_1, I_2 are equal when the legs are fully extended at right angles to the body. The photographs* given in Plate II. show that it first contracts its fore legs (thereby making I_1 less than I_2) and then turns its fore part round. This latter action necessitates the hind part being turned in the *opposite direction* (since the total angular momentum about the axis is zero) but to a *less extent*, since I_2 is $> I_1$. The animal then contracts its hind legs, extends its fore legs, and gives its *hind part* a turn. This necessitates the fore part being turned in the reverse direction but, again, to a less extent, since I_1 is now $> I_2$. It will thus be seen that by continued action of this kind the cat can turn itself through any required angle, though at no time has it any angular momentum about its "axis."

[23.] If the cat were allowed to fall on to a cushion, resting on a smooth floor, would the fall of the cat move the cushion ?

57. The foregoing remarks may be summed up in the following two principles connected with the motion of a gyroscope such as we have been considering :

1. *Since there is no torque about a vertical axis, the angular momentum about the vertical through the point of support must throughout the motion remain constant, i.e. equal to the value, whatever it was, at the beginning of the motion.*

2. *The torque, which is about a horizontal axis, must produce angular momentum about that axis, and it cannot produce angular momentum about any other axis.*

These are of the utmost importance in considering the motion of a gyroscope.

An explanation of the phenomenon of precession by the principle of conservation of angular momentum will be found in the Appendix.

58. Why precession increases as a top dies. It will be remembered that if a gyroscope is spun with one end of its axis in the socket of the vertical stand, Fig. XVI., then towards the end of the motion it precesses faster and faster, as it descends, until it finally flies off the stand.

Let us apply the principle of conservation of angular momentum to the consideration of this phenomenon.

We will suppose the wheel to be spinning initially with angular momentum $I\omega$, its axis being inclined at an acute angle θ_0 to the vertical. If we neglect all friction,† then the

---

* These photographs are reproduced from *Nature* (Nov. 22nd, 1894), by the kind permission of the proprietors.

† In reality the friction cannot be entirely neglected. The friction at the socket resists precession, and that at the bearings of the wheel causes a gradual diminution of its spin, but these retardations are not enough to affect the statement that the precession increases as the gyroscope descends. A further discussion, taking some of the friction into account, is given in Art. 66.

angular momentum about the vertical through the point of support must remain constant throughout the motion, *i.e.* equal to the initial value $I\omega \cos \theta_0$. Precession will take place in the *same* direction as ω viewed, say, from above. As the gyroscope descends, its inclination θ to the vertical increases, and therefore $I\omega \cos \theta$ diminishes. [Or the projection of the disc as seen from above diminishes: see Art. 31.] Hence the contribution of angular momentum about the vertical (in the *same* direction), due to the precession of the frame and wheel (regarded as not spinning), must increase, in order to maintain the constancy of the whole.

When the axle becomes horizontal it is clear that the whole angular momentum about the vertical through the point of support is due to the *precession* of the wheel and frame, and if I_z is the moment of inertia about the vertical for the frame and wheel, and Ω the (increasing) azimuthal velocity, we have

$$I_z\Omega = I\omega \cos \theta_0,$$

$$\text{or} \quad \Omega = \frac{I\omega \cos \theta_0}{I_z}.$$

After this the contribution of angular momentum about the vertical from the spinning of the wheel is in the *opposite* direction to precession ($I\omega \cos \theta$ has become negative), and since the constancy of the whole is maintained, the angular momentum about the vertical due to precession must continue to increase.

This last position of the gyroscope corresponds to the top represented in Fig. III. (*a*) of the Introductory Chapter.

The same considerations hold good in the case of a top spinning on a fine fixed point, except that in this case the top touches the table before θ reaches 90°.

59. Inertia. We have already alluded in Art. 8, to the resistance due to the inertia of a body, *i.e.* the resistance which a solid body offers, by reason of its mass, to any force applied to it. If the body had no mass it would offer no resistance, provided it were free to move in the direction of the applied force.

The following conception of a solid body is frequently useful in illustrating the property of its inertia.

Suppose that we imagine a perfectly rigid massless skeleton framework, free to move in every direction, and composed of a great number of small hollow cells with smooth thin walls. Since it is perfectly massless it is incapable of offering resistance to force. Next suppose that each cell is filled with a loosely fitting shot, as in Fig. 28.

The whole body has now considerable inertia, due to the mass of the shot which are in the cells: and we may consider a rigid body as composed of an infinite number of small heavy particles (corresponding to the shot in the cells) kept in proximity to one another by a massless framework similar to the one above described.

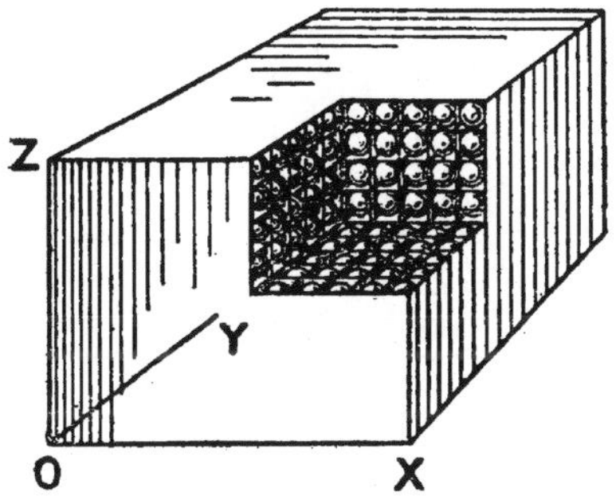

FIG. 28.

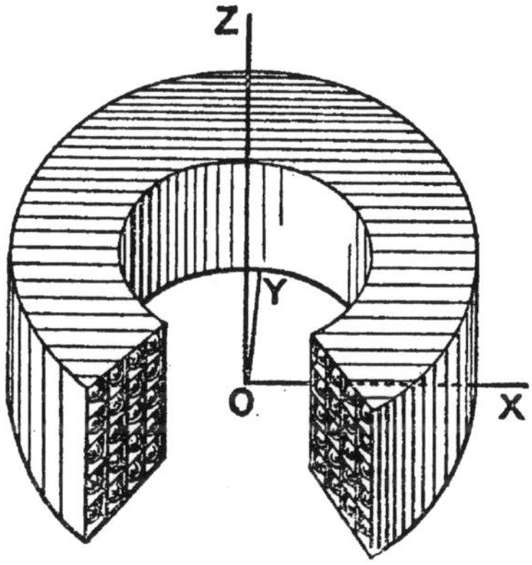

FIG. 29.

We know that a rigid body resists force in *every* direction; or, to follow out our analogy, in whatever direction our framework is pushed, a resistance will be experienced, since the moment the framework moves, each shot begins to press against a side of its cell. Let us suppose however that we could conceive an imaginary body which possessed inertia in two directions, say OX and OZ, but not in the direction OY, such a conception might be approximately illustrated by considering the skeleton framework to be composed of long *tubular* cells, running parallel to OY: for in this case an attempt to push the *framework* in the direction OY would meet with no resistance. Again, if the tubes were *circular* in shape as in Fig. 29, and endless, their centres lying on the line OZ, then the body would offer *no resistance* to being *turned about* OZ.

60. Start of precession. In Art. 34 we explained that the equation $K=I\omega\Omega$ applies only to the maintenance of existing precession and not to the starting of precession. We have yet to consider what happens when precession is first set up.

Let a weight be attached to the gyroscope as in Figs. 30 and 31. This will clearly introduce an external couple which acts on the gyroscope, but it should be noticed that there is one already, due to the weight of the wheel and frame. The additional weight will, however, give a clearer illustration.

Suppose that the wheel is spun in the direction indicated in Figs. 30 and 31 while its axle XO is *held* in a horizontal position, and let it be then released so that the attached weight and the weight of the frame and wheel begin to act. It will be observed that the whole system "wobbles" and oscillates considerably, and then finally settles down to a steady precession about the vertical OZ.

Let us consider the reason for these phenomena, assuming the spin of the wheel as constant throughout the motion. When the gyroscope is released, the axle tends to precess about *OZ* under the action of the external couple; but, before it can do so, both the wheel and frame and also the attached weight have to acquire angular momentum about *OZ*, about which there is no external torque. What we may call the inertia "about *OZ*" acts as a resistance to the precession, and *causes the wheel to dip downwards* (Art. 47) through a small angle. Resistance due to inertia is clearly seen in the case of the attached weight which will "lag behind."

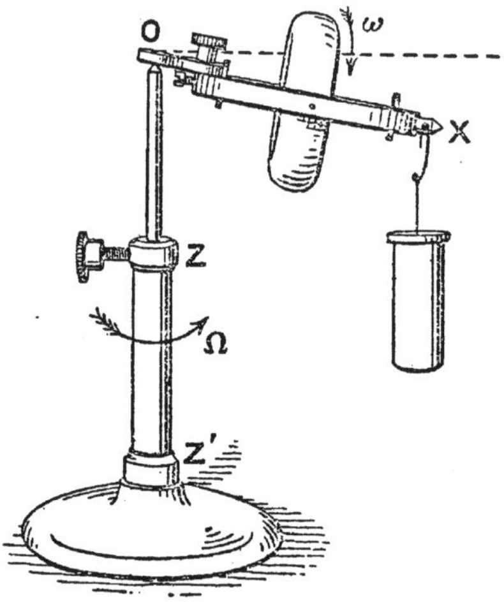

FIG. 30.

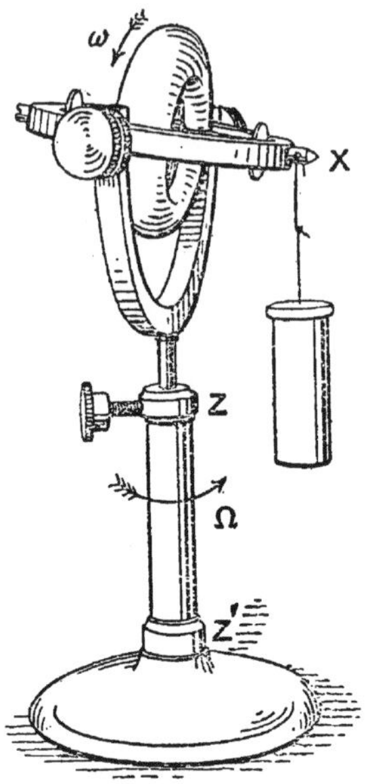

FIG. 31.

FIGS. 30 and 31.—Gyroscope when released, inclined at a small angle to the horizontal. The angle in the figure is somewhat exaggerated for the sake of clearness.

Now when *OX* dips down out of the horizontal, an observer with his eye vertically above *O* (Art. 31) sees a projection of the disc of the wheel; so that it is clear that as soon as the axle dips, the *spin of the wheel* contributes angular momentum about *ZO* (not about *OZ*). But since there is no torque about the vertical *ZO*, the angular momentum of the whole system about this vertical must always be zero, since it was zero originally. Hence, as the gyroscope dips, the frame, weight and wheel must acquire angular momentum about *OZ* (not about *ZO*) by the motion of the centre of gravity round *OZ*. This will cancel that of the wheel about *ZO*, due to its spin about its axle, and maintain the constant total value zero. In other words the wheel, frame, and weight will precess about *OZ*.

61. The above phenomenon of the dipping of the gyroscope illustrates also the second principle laid down in Art. 57, *i.e.* the torque can only produce angular momentum about its own axis. For it is clear that when the system is precessing its kinetic energy is greater than before it was released; since, as far as the spin of the wheel is concerned, it is the same, and the wheel, frame, and weight are now precessing. The fact is, this excess of kinetic energy is provided by the work done by the torque *in turning the system about its own axis,* and causing it to dip. Hence the dip is necessary for conservation of energy as well as for conservation of angular momentum.

62. Oscillations. This precession will not *at first* be steady: for we have seen that whether precession is taking place steadily or not, the angular momentum about the vertical of the frame, weight, and wheel (regarded as not spinning) must be cancelled by the angular momentum about the vertical due to the spin of the wheel—which necessitates the dipping of the axle.* Hence the precessional velocity depends on the dip, and alters if the dip is altering. There will however be a steady value of Ω, consistent with $I\omega$ and K, and also a steady angle at which the axle dips to the horizontal: and this steady or mean value of the "dip" and of the precessional velocity Ω, is arrived at in the following manner by a series of impulsive jerks and oscillations which are eventually destroyed by friction.

The particles which at first, by their inertia, retard the precession, as soon as "dip" takes place acquire, by impulsive action, a velocity of precession greater than the mean velocity Ω of the axle, and now, owing to their momentum, have the effect of hurrying the precession. The next instant the precessional velocity of the axle will be greater than that of the particles, and their inertia will again retard precession; and so on.† But every time the precession of the axle is hurried or retarded the centre of gravity of the system rises or falls (Art. 47). The result is that, in addition to the backward and forward "wobbling" the gyroscope oscillates up and down about the mean position of "dip," until finally, the oscillations having

---

* By hurrying or retarding the precession the angle made with the horizontal by the axle of the gyroscope can be diminished or increased: but if the system is *left to itself* it must precess with a small "dip."

Although the axle is precessing at an inclination to the horizontal we may still employ the equation $K = I\omega\Omega$ since the angle in question is small. In Chapter VI. this equation is discussed for all inclinations.

† The successive retarding and hurrying effect of the inertia can here again be most clearly realised by observing the suspended weight, which swings backwards and forwards in jerks about a mean position.

An analogy to this starting of precession may be found in the motion of two railway trucks coupled together at rest, when one receives a sudden impulse. Each in turn assists and hinders the other's motion, until the same steady motion is arrived at for both.

been destroyed by friction, an angular velocity Ω is reached consistent with the values of K and $I\omega$, *i.e.*, when

$$\Omega = \frac{K}{I\omega}.$$

Steady motion will then follow; but any attempt to alter this existing state of steady precession will be followed by oscillations similar to those observed at the starting of precession.

It should be noticed that during the motion above described, the gyroscopic resistance $I\omega\Omega$ (Art. 45) is alternately greater and less than the applied couple.

It is clear, then, that the "dip" and consequent oscillations are entirely due to what we may call the inertia "about OZ" of the frame, attached weight, and wheel; but if they possessed the imaginary property of having no inertia "about OZ" as suggested in Art. 59, there would be no dip and no oscillations. In fact the gyroscope would precess exactly as it does in the case of steady motion where there is no inertia to be overcome. The reactions at the bearings of the wheel contribute slightly to the oscillations, in the same way that the inertia does, but are not the primary cause. The primary cause of the oscillations is the inertia of the system.

63. The reason will now be seen for the phenomenon mentioned in the Introductory Chapter, that if a downward pressure is applied at X (Fig. XXII.), when the gyroscope is spinning slowly, the axle XX' dips appreciably before precession takes place, but a sudden removal of the pressure will leave the system oscillating violently: while if the spin is very fast, the dip is hardly perceptible, and the oscillations are scarcely more than a "shiver." For, in order that steady motion may be arrived at, Ω $\left(\text{which equals } \frac{K}{I\omega}\right)$ must be large when ω is small, and consequently the impulsive actions and reactions due to the inertia will be large before the required value of Ω is reached; whereas, when ω is large, Ω is small, and in this case the dip and oscillations will be small also.*

A similar explanation to the above accounts for the oscillations or *nutations* of a top. These are especially conspicuous just before its fall, when it goes through a sort of reeling motion. The spin has by this time been considerably diminished, and

* The reason for this may also be partly seen from the parallelogram of velocities. For, in the accompanying figure, if ω' represents the velocity created

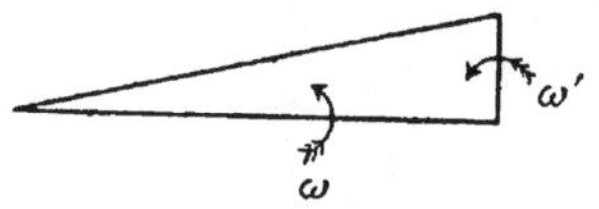

by the applied couple about its own axis, it is clear that when ω is large the axis of resultant angular velocity is only slightly displaced; and *vice versâ*.

consequently the friction which in the earlier part of the motion tended to "damp" the oscillations.

64. The above investigations may now be summed up as follows:

Whenever we apply to the axle of a spinning gyrostat a torque which *does more than maintain existing precession*, this torque produces three results:

(α) **Oscillations are set up and continue until destroyed by friction. They may be so quickly destroyed, or so slight, that they are hardly appreciable to the eye.**

(β) **The torque produces, about its own axis, an effect which is not always appreciable to the eye, and corresponds to the dipping of the gyroscope in the preceding articles. Thus we see that work is done on the gyrostat.**

We shall in future pages allude to this effect of a torque about *its own axis* as "dip."

(γ) **It produces about an axis perpendicular to itself and to the axle, an effect which is much more appreciable to the eye, and corresponds to precession. Thus we realise the increased kinetic energy due to the work done on the gyrostat.**

This effect of the torque we shall allude to as precession.

Lastly, it must be remembered that there can be no change of angular momentum about an axis round which no torque has acted.

65. Explanation by the principle of energy of "Hurry the precession, the top rises." We shall now see more clearly (compare Art. 47) why a spinning top rises if we hurry its precession. For we have seen that the "hurrying" force produces only a small hurrying effect on the precession (corresponding* to "dip"), but the turning effect (corresponding to precession) is much more appreciable, and raises the top. We are, of course, doing work on the top by hurrying the precession, and since the kinetic energy of precession is not appreciably increased, the potential energy of the top must be increased, *i.e.* the centre of gravity of the top rises. In the case when friction at the toe alone hurries the precession, the work which raises the top is done at the expense of the kinetic energy of spin of the top.

When precession is retarded, work is done by the top, which, since it does not lose much kinetic energy of precession, loses potential energy, *i.e.* the centre of gravity descends.

66. Explanation by the principle of energy of the increase of precession in a dying top. As in Art. 58, let us consider that a gyroscope is placed in the socket of its stand with its axis at an angle θ_0 to the vertical, when the angular momentum

* As the spin of the top diminishes the hurrying effect is greater. See Art. 63.

about its axis is $I\omega$. Then $I\omega \cos \theta_0$ is the angular momentum about the vertical through the centre of the socket, and this angular momentum would remain constant throughout the motion, but for the diminution owing to the frictional couple at the socket *on the frame.* We can neglect the effect of friction on the spin of the wheel, for it will be small.

Now we have seen that the work done by the couple due to gravity

$$-\text{K.E. of precession}$$
$$+\text{K.E. of descent or ascent (\textit{i.e.} ``dip''),}$$

and since the latter of these two contributions is not large, it is clear that most of the work done shows itself in K.E. of precession, and for this reason the gyroscope precesses faster as it falls.

Again, the work done by the frictional couple at the socket

= K.E. of descent (corresponding to precession)
+ K.E. taken from existing precession (corresponding to "dip").

Of these two contributions the former is the greater, and therefore the gyroscope also *descends* more quickly as it falls.

The same principles govern the motion of an ordinary spinning top which is gradually falling from a vertical position, or one which has never risen to the vertical, but is descending from its position of steady motion.

The gravity couple, by a large number of excessively small "dips," gradually causes the top to fall, and, simultaneously with this, produces the more appreciable increase of precession. The friction couple at the toe (besides destroying the spin of the top) resists the increasing precession—the result, about its own axis, corresponding to "dip"—but contributes, more appreciably, to the falling of the top—the result corresponding to precession. A further discussion of this motion is to be found in Chapters VII. and IX.

If the toe of the top is free to move instead of being constrained as in the above instance, the same principles apply; for we can consider the centre of gravity of the top as a fixed point for purposes of rotation, and discuss its translational motion independently.

67. *To prove that when the axle of the gyrostat has been released from its horizontal position, as much energy is lost as is communicated to the system.*

Let the axle (Fig. 30) on being released descend, by oscillations, through an angle $\delta\theta$ before steady motion is reached. Let the angular momentum about the axle be $I\omega$, supposed constant throughout: we shall however take into account the friction at O, which assists in destroying the oscillations. Let I_z be the moment of inertia about OZ of the frame and wheel, and Ω the final velocity of precession.

Now the angular momentum about the vertical due to the dipping of the wheel is $I\omega \sin \delta\theta$, or $I\omega . \delta\theta$, since $\delta\theta$ is small. The angular momentum about the vertical due to precession is $I_z\Omega$; and we have seen that these two are equal in value but opposite in direction.

Hence, $$I\omega . \delta\theta = I_z\Omega. \quad\text{.............................(i)}$$

Again, the kinetic energy communicated

$$= \tfrac{1}{2} I_z \Omega^2$$

$$- \tfrac{1}{2} I\omega\Omega\delta\theta, \quad \text{by (i)}.$$

But the whole work done

$$= K\delta\theta$$

$$= I\omega\Omega\delta\theta \cdot$$

therefore half the work done, *i.e.* half of the energy communicated, is lost in frictional heat.

It should be noticed that the couple K is not *quite* constant throughout the period under discussion, since it varies between K and $K \cos \phi$, where ϕ is the inclination of the axle to the horizontal at any particular instant: but since ϕ never differs much from $\delta\theta$, the variations in K are negligible.

Analogous theorems are known in Electricity and Magnetism. Lord Kelvin first pointed out that if the currents of a system are kept constant by a battery during a displacement in which the electro-kinetic energy is increased by W, an equal quantity W of energy is dissipated from the wires as heat, the battery being thus drawn upon for a double quantity of energy $2W$. See Maxwell's *Electricity and Magnetism*, 3rd Edition, Vol. II., p. 225. See also Vol. I., p. 120, for a similar theory in Electrostatics.

CHAPTER V.

PRACTICAL APPLICATIONS.

I. THE STEERING OF A TORPEDO.

68. An allusion has already been made on p. 29 to the gyroscopic mechanism employed for automatically steering a torpedo. The patent was originally protected by M. Obry, an Austrian engineer, and sold to the authorities of the Whitehead Torpedo Works at Fiume, by whom it was improved and finally patented in its present form in 1898. The following is a brief outline of the way in which the steering is effected.

As the torpedo passes through the impulse tube, a trigger projecting from its upper surface catches against a bolt in the tube and releases a spring by which the gyroscope is spun (see p. 29). Thus, before the torpedo enters the water, the axle of the gyroscope is pointing in the required direction, from which it never deviates (p. 11). Any deviation sideways on the part of the torpedo only alters the position of the gyroscope *relatively to the torpedo.* The gyroscope is fitted to the torpedo in such a way that this relative change of position opens one of two valves (the other being temporarily closed) connected with the compressed air chamber from which the screws of the torpedo are driven.

The air rushing through either valve drives (in one case forwards and in the other backwards) a piston rod connected with a vertical rudder at the stern of the torpedo, and the valves are so arranged that when the torpedo swerves to starboard the rudder steers to port, and *vice versâ.* The middle position when both valves are closed is scarcely ever maintained for an appreciable time, with the result that though the torpedo maintains the direction intended it is by means of a zig-zag path, roughly 2 ft. broad.

The mechanism is in addition so arranged that if the position of the object fired at prevents the torpedo being aimed directly at it, it is possible to set the gyroscope (apart from the torpedo) in such a direction that the latter will eventually strike the object, although in the first instance not aimed directly at it.

Full particulars and diagrams of the invention can be procured at the Patent Office, Southampton Buildings, London.

II. SCHLICK'S METHOD OF STEADYING VESSELS AT SEA.

69. Herr Otto Schlick has recently carried out a series of valuable experiments in the "See-bar," formerly a first-class torpedo-boat of the German Navy, with a view to applying the principle of the gyroscope to counteract the rolling of a vessel at sea. We have seen (p. 42, question 11), that when a paddle steamer heels over to starboard, either by the action of a wave or from some other cause, her bows, owing to gyroscopic action, turn towards the starboard side, and *vice versâ*. Consequently, a paddle steamer rolls less than an ordinary screw steamer, though the gain in this respect is somewhat at the expense of the *directness* of her course. But it will be noticed that when she heels to starboard, the starboard wheel, by dipping further into the water, exerts more power than the port wheel, and so tends to turn the bows to the port side. Thus the gyroscopic effect of the paddle-wheels tending to steer the vessel to starboard meets with a counteracting tendency from the starboard wheel to steer the vessel to port. Otherwise it would be still more difficult than it now is to keep a straight course with a paddle steamer in stormy weather.

If a large fly-wheel were mounted in the middle of a screw steamer on a horizontal axle at right angles to the length of the ship, and were made to revolve rapidly, it is clear that the steamer would become much steadier, but only in so far as she was allowed to make deflections from her course. To obviate this difficulty Herr Schlick places the fly-wheel with its axis vertical, and mounted in a frame (see Plate III.) which can itself turn about a horizontal axle directly athwart ship. The rolling force of the waves is thus counteracted (in part at any rate) by the gyroscopic resistance of the rotating wheel, while the wheel itself turns about the horizontal axle of the frame, but in no way interferes with the direction of the ship's course. Two effects are thus brought about:

(i) the rolling force of the waves is resisted;

(ii) the period of the oscillations is consequently lengthened, so that it no longer synchronizes with the motion of the waves.

The latter will thus tend to damp the vessel's oscillations, for it should be remembered that violent rolling is the accumulated effect of a series of waves each adding to the existing rolling.

The oscillations of the fly-wheel itself are discussed in the next article.

When the above appliance was first suggested many seamen expressed the opinion that waves would more easily break over a vessel steadied in this manner than if it were free to take up the motion of the sea, but experience has proved that this is not the case. A *vertical heaving motion* takes place, but the tendency of waves to break over the vessel is less. During experiments conducted on July 17, 1906, the "See-bar" when broadside on to the sea, *with gyroscope fixed*, reached a maximum inclination of 25° to leeward and 15° to windward. Her rolling motion passed through "phases" gradually increasing and then decreasing. Not long before the gyroscope was set free the rolling attained 15° on each side of the vertical—an arc of rolling of 30°; but immediately the gyroscope came into operation the arc of rolling (out to out) was reduced to 1°.

70. The following detailed account of experiments leading to the more technical applications of the above results is reprinted by kind permission of the Institution of Naval Architects, being an extract from a paper read before the society by Herr Schlick on March 24, 1904.

"The phenomena which present themselves in connection with an arrangement of this kind may best be studied by the help of a model, such as that illustrated in Fig. 32. This shows a pendulum which is able to swing to and fro on an axis at n. Above, the pendulum takes the form of a semi-circular fork-piece B. A ring is hung on the point of the screw-pins pp in such a manner that it can turn about a horizontal axis through the centre of the fork-piece. In the ring itself, which is so weighted that, when inclined by any means, it will always return to the upright position, a vertical spindle, carrying the fly-wheel F, is set in bearings. If the pendulum be set swinging without the fly-wheel being made to rotate, it will be found to move to and fro with a certain definite period. The frame in which the fly-wheel is borne may first be allowed to move as easily as possible, *i.e.*, with a minimum of friction. If the fly-wheel be then set spinning and the pendulum receive a push, the latter will, in the first place, show a considerably increased period of swing.

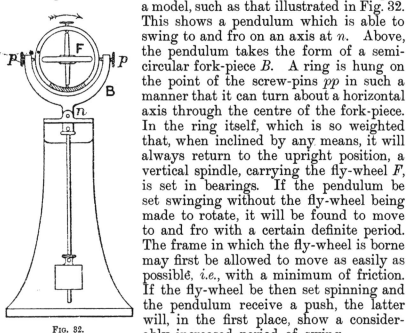

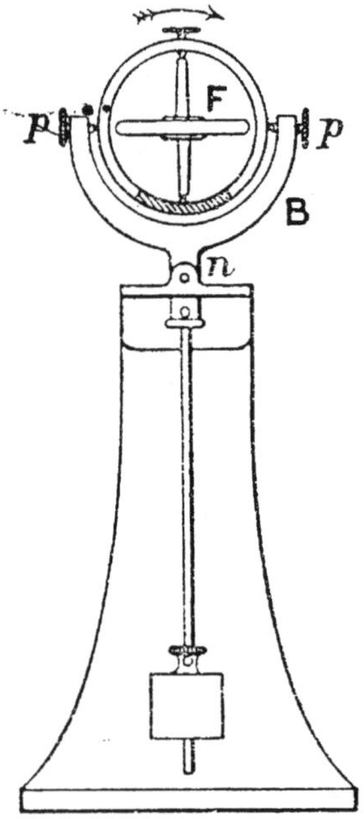

Fig. 32.

"The fly-wheel oscillates with its frame during the swings of the pendulum with a so-called phase difference of 90°, that is to say, the swings of the axis of the fly-wheel keep lagging behind those of the pendulum by a quarter of a swing. While the pendulum is passing its central (vertical) position, the axis of the fly-wheel will show its greatest inclination, and when the pendulum is in the outermost position of its swing, the axis of the fly-wheel stands exactly in its middle position. The amplitude, or extent of the swing, of the pendulum will, as theory tells us, not be influenced thereby, but will remain exactly as great as before. That this is the case will also be readily apparent, since no consumption of energy takes place in the apparatus, the period only being influenced by the increase which takes place in the swinging mass.

"Were it possible to fit a fly-wheel of this kind, able to swing in its frame without experiencing friction, into a vessel, this would be advantageous in so far that, to begin with, the rolling motions would become slower, and therefore less unpleasant, and then, on account of the great difference thus produced between their period and that of the waves, they would cease to be of any consequence. The rolling motions of the vessel would then become considerably less in extent. If the frame which bears the fly-wheel be screwed tight on the model, so that it can no longer turn, the effect hitherto produced by it will cease, and the pendulum will swing with the same period as it would if the fly-wheel were not rotating.

"It will readily be seen that the effect produced upon the swings of the pendulum by the rotating fly-wheel can be of greater extent only so long as the plane of the frame bearing the fly-wheel remains approximately vertical.

"If the axis of the fly-wheel be inclined at an angle α to the vertical, the moment thus produced, acting against the motion of the pendulum, will be proportional to the value of $\cos \alpha$. Should the axis of the fly-wheel momentarily become horizontal, a position which with a pendulum in violent motion it may almost reach, that is to say, should $\alpha = 90°$ and $\cos \alpha = 0$, the influence of the fly-wheel will disappear altogether.

"Since, as already stated, there is a phase difference of 90° between the swings of the pendulum and those of the axis of the fly-wheel, the gyroscopic influence on the pendulum must be least in amount when it is passing the middle position, *i.e.*, at the very position at which it has its greatest angular velocity, because at the same moment the inclination of the axis of the fly-wheel is at its greatest, while the velocity with which it is changing its inclination has become very small or vanished altogether. When, on the other hand, the pendulum has reached its outermost position, and is changing the direction of its motion, thus for an instant reaching a state of rest, the axis of the fly-wheel then proceeds with its greatest angular velocity through the middle position, the fly-wheel thereby exerting its greatest influence. It will thus be evident that the conditions for the exertion of the greatest possible influence of the gyroscopic action on the pendulum are not present here. In order that the motion of the pendulum may be effectively influenced, the oscillation of the frame with the axis of the fly-wheel will have to be reduced in a suitable degree. In the model illustrated in Fig. 32 this may be most simply effected by tightening the screws *pp* to a suitable extent, so that they act as a brake on the motion of the frame of the fly-wheel. The swings of the fly-wheel frame are thus reduced in extent, and the phase difference between the two swinging movements here described now becomes less than 90°.

"If the experiment be made of setting the pendulum swinging with the brake thus applied to the fly-wheel frame, a very different phenomenon will be observed. The pendulum will indeed still swing with a very considerable period, but the maximum angle attained becomes considerably reduced with each successive swing, so that a state of rest is reached even after about two complete swings. In scientific language, the oscillations of the pendulum experience a damping, in that the energy stored up in it is converted into heat by the friction applied to the fly-wheel frame."

The complete paper from which the above is an extract can be procured at 5 Adelphi Terrace, London, W.C.

71. The first gyroscopic apparatus for steadying ships which was constructed in England, was made at Newcastle, at the Neptune Works of Swan, Hunter and Wigham Richardson, and was fitted in October, 1908, to the R.M.S. "Lochiel," owned by Messrs. David MacBrayne of Glasgow. It can be thrown in and out of action at will. When it is out of action, the vessel has been observed to roll, out and out, through an arc of 32 degrees, which was reduced to a (total) angle of from 2 to 4 degrees by the action of the gyroscope. The machinery, which occupies very little space in the steamer, is driven electrically and requires very little attention.

III. BRENNAN'S MONORAIL.

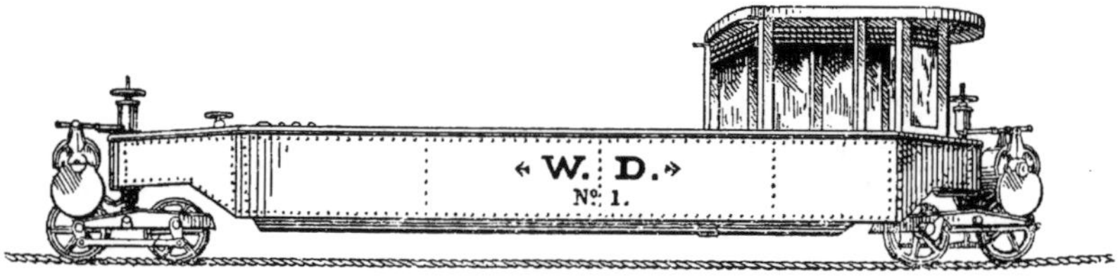

Model exhibited before the Royal Society, May 8, 1907.

72. The gyroscope has recently been employed by Mr. Louis Brennan with striking ingenuity and success to ensure the stability of a heavy car travelling on a single line of rail with its centre of gravity *above* the level of the rail, as is seen in the accompanying illustrations.

The following pages will serve to illustrate the development and growth of his idea from the first elementary principles; but it should be remembered that the application of the gyroscope to methods of locomotion is still in its infancy, and consequently the details of construction in Mr. Brennan's Monorail in its final form will probably differ largely from those which are here explained.

The invention,* in its simplest form, consists in affixing to the car a heavy-rimmed fly-wheel, or gyrostat, *AB*, revolving in the same plane as the road wheels and in the same direction

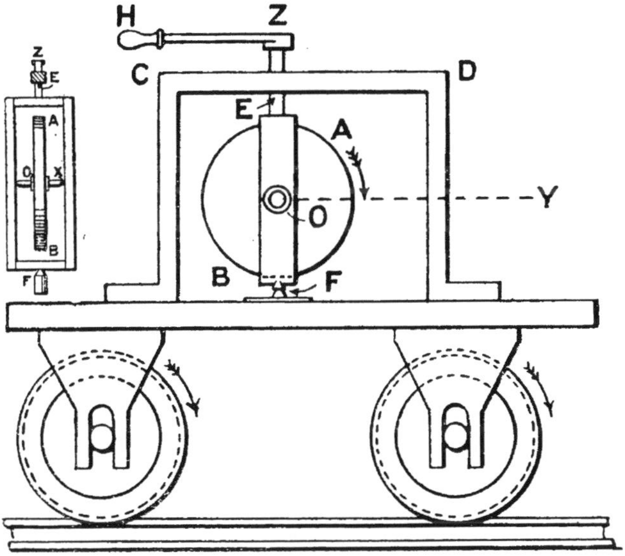

FIG. 33.

(Fig. 33), on an axle *OX* perpendicular to the plane of the paper. If it were drawn, *X* would be towards the reader. This axle is mounted in a frame or carrier *EF*, which is pivoted at *E* and *F* so that it can turn about an axis *OZ* coinciding with *EF* and perpendicular to the axle of the fly-wheel. This turning can be effected either by muscular power applied to the handle *H*, or by some automatic arrangement.

Suppose that when the car is travelling in a straight line, some force, such as wind pressure, causes it to lean out of the vertical, then, a tilting couple having been applied to the axle of the wheel, the wheel begins to precess about the vertical *OZ*.

But, as in the case of the top, if we "hurry the precession, the top rises," so, if the precession about *OZ* is hurried, either by applying a force at the handle *H* or by the automatic control, the fly-wheel will immediately rise to a more vertical position carrying the car with it, and so will restore the vertical position of the car. The control of the precession might be effected by the automatic action of an inverted pendulum.

*The provisional and complete specifications of the Patent, from which this description is taken, can be procured from Messrs. Marks & Clerk, 18 Southampton Buildings, London, W.C. Price 8d.

Now let us consider what happens as the vehicle rounds a corner to the right, say. As the vehicle turns, the gyrostat maintains its plane of rotation unaltered in direction, and is thus, *relative to the car*, displaced out of its central position in the plane *FCD*. To obviate this displacement, Mr. Brennan makes use of a second gyrostat exactly similar to *AB*, mounted (preferably) in the same plane on a parallel axle, but *rotating in the opposite direction*; and the carriers of the two gyrostats are connected by means of gearing, so that the rotation of one carrier in one direction ensures a corresponding rotation of the other carrier in the opposite direction. The movements of the carriers are then controlled as before by a lever or other suitable means. The gyrostat to which the device for accelerating precession is fixed is called the actuating gyrostat. Both are made to rotate *in vacuo*; otherwise the resistance of the air at the necessary high speed of rotation would be extremely great.

It will be noticed that the above principles apply whether the car be moving backwards or forwards: also that the centre of gravity of the vehicle can be made to move laterally in relation to the axis of support without bringing gyroscopic action into play, and thus the car is enabled to move round a curve while maintaining a vertical position.

73. In order to illustrate the stability of Mr. Brennan's car, Messrs. Newton of Fleet Street, London, make a gyroscope

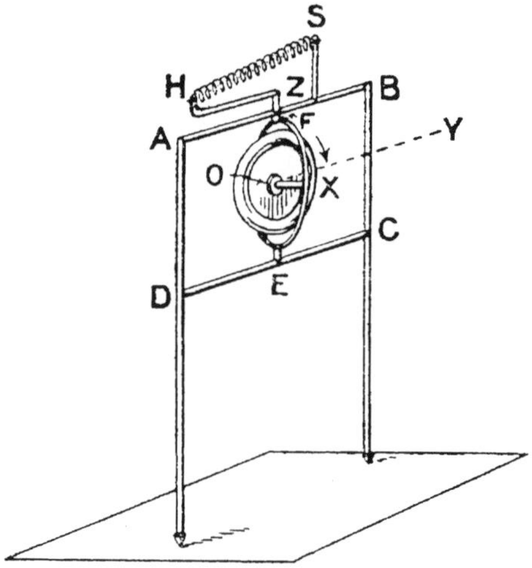

FIG. 34.

mounted as in Fig. 34, which will be seen to be virtually the same as a model of the invention in its simplest form, Fig. 33. In Fig. 34 an addition has been made to what (at the time of writing) is actually sold by Messrs. Newton, suggested by

Professor H. A. Wilson and exhibited by him at a meeting of the Physical Society of London in November 1907, as a method of automatically hurrying the precession. It consists of a spiral spring *SH*, one end of which, *S*, is fixed to the rectangular frame *ABCD*, while the other end, *H*, is attached to the small crank *ZH*, which turns about *OZ* as the wheel precesses, and is rigidly fixed to the carrier *EXF*. It will be seen that so long as the wheel, and consequently the crank, are in the plane of the rectangular frame *BD*, the tension in the spring has no moment about *OZ*; but immediately the wheel precesses and the crank moves out of the plane of the frame, the tension of the spring has a moment which tends to hurry the precession, and consequently causes the wheel and frame to return to their vertical position. But the momentum of the wheel and frame will cause them to overshoot the vertical position,* so that the wheel will again precess out of the plane of the frame, the precession will again be hurried and the previous process will be repeated. Thus oscillations are set up of increasing amplitude, with the result that the frame and gyroscope eventually fall over. Professor Wilson thus illustrates the necessity for "damping" the oscillations set up by any automatic hurrying of the precession. We understand that Mr. Brennan has several means which he employs in order to damp these oscillations, one of which is clearly explained in the following article from the *Times* Engineering Supplement for June 5th, 1907, which is here reproduced by the kind permission of the author, Professor Worthington, late Headmaster of Devonport Naval Engineering College, and that of the *Times* authorities. Apart from its immediate bearing on the monorail, the article serves as an excellent recapitulation of much that has already preceded it in previous chapters.

"THE BRENNAN MONO-RAIL CAR."

"The interest excited by Mr. Brennan's mono-rail car, both among the general public and among engineers, is due partly to the obvious importance of the possible revolution in methods of transport which his car suggests and partly to surprise at the results he has obtained from what we believe is an entirely new application of gyrostats in combination.

"The majority even of mechanical engineers, to say nothing of the general public, have too little experience of the properties of the gyrostat to have been able to realize what it is that Mr. Brennan has done, or to appraise his inventions. We say inventions, rather than invention, for the stability of his car depends, as his patent specifications show, on at least three

* It will be found that owing to this tendency on the part of the gyroscope to overshoot the central position, we can, by a succession of taps on one side of it, cause it to fall over on that side.

distinct inventions—(1) the automatic calling into play of a force tending to accelerate precession, by the rubbing of the axle of spin as it rolls along a guide, an action which may be said to be borrowed from a spinning peg-top; (2) the regulation of the precession so as to leave the gyrostats, after any disturbance, always with their planes of spin parallel to the rail; (3) the combination of two linked gyrostats spinning in opposite directions, for meeting the exigencies of a curved track.

"To understand these points it is necessary to recall the behaviour of Foucault's gyrostat. Let the diagram (Fig. 35) represent a heavy-rimmed disc with its spindle *OA* horizontal.

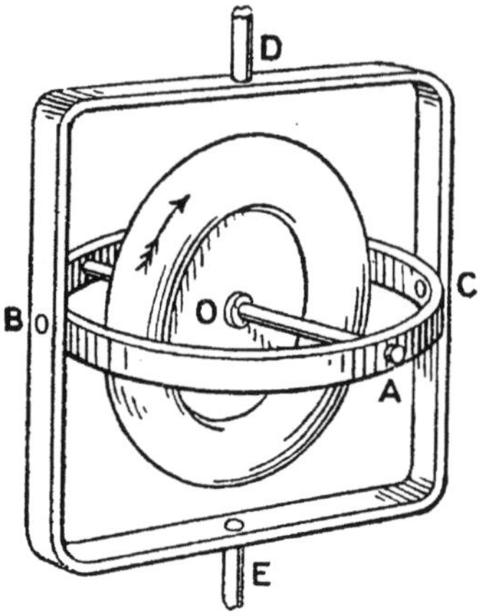

FIG. 35.

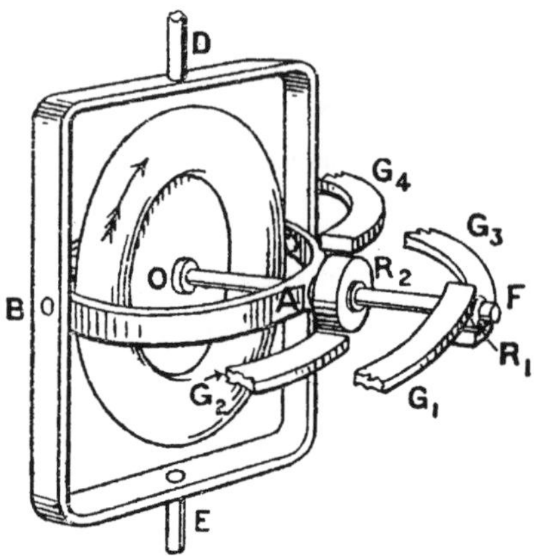

FIG. 36.

This spindle turns in a ring or frame *BAC*, pivoted about the horizontal axis *BC*, at right angles to *OA*, on the frame *BDCE*, which frame again is free to turn about the vertical axis *DE*. If the disc be not spinning, a downward pressure at *A* will cause *A* to descend with acceleration. If the disc be spinning slowly in the direction *BDC*, shown by the arrow, then *A* will not merely descend, but will move towards *B*, with rotation about *ED*, and a sudden removal of the pressure will leave the whole system violently oscillating.

"If, however, the disc be spinning very fast, then we find that a downward pressure at *A*, maintained constant, produces no sensible depression of *A*, but creates what is called a precessional rotation of the spindle *OA*, at a constant rate about the axle *DE*, *A* moving towards *B* along the arc *CAB*, at a constant rate so long as the rate of spin is unaltered. When the pressure ceases, there is only a slight and perhaps imperceptible tremor of the spindle *OA*, and we are left with the spindle displaced horizontally through an angle which is a measure of the time-integral of the couple that has been applied about the axis *BC* as it moved.

"If, on the other hand, with the disc spinning very fast, the pressure maintained at *A* is not vertical but horizontal, in the direction of the precessional rotation just described, then there will be no apparent revolution about *ED*, but there will be a rotation of the frame *BAC* about *BC*, and *A* will rise, the angle through which *OA* rises being in this case a measure of the time-integral of the moment of the pressure about *ED*.

"If now, while we maintain a constant downward pressure at *A*, we also apply a constant horizontal pressure as if to accelerate the horizontal precession of the spindle, then the precession will not indeed be permanently appreciably accelerated, but *A* will rise at a rate proportional to the horizontal force, and work will be done by the horizontal and against the vertical force.

"These are the chief relevant physical facts which lay at Mr. Brennan's disposal and of which he has availed himself with such remarkable skill and success. We will endeavour to explain his arrangement by pointing out its relation to the Foucault gyrostat just described. In the first place the frame *EBCD* is pivoted on the body of the car at *E* and *D*, so that when the car is erect *DE* is vertical. Mr. Brennan then makes the spindle of his disc into the armature of an electro-motor, whose field-magnets are carried by the pivoted and still balanced frame *BAC*. When everything is in equilibrium and the car is running erect, the spindle *OA* is horizontal and at right angles to the rail. We will suppose the car to be running in the direction *BC* and will refer to *A* as the right-hand end of the spindle.

"Now let a wind-pressure be applied to the left side of the car. The car begins to turn over relatively to the gyrostat, and thus at once brings down a guide plate G_1 (see Fig. 36) fixed to the car, and bent into a circular arc as shown, so as to press on a small roller R_1, turning loosely about the end *F* of the spindle *OAF*, which now projects beyond the frame *BAC*. The pressure on this roller causes the spindle to precess from *A* towards *B* with an angular velocity proportional to the pressure, and the turning-over of the car is arrested, but at the same moment the friction between the rotating spindle and the roller makes the latter roll along the under side of the guide-plate and thus evokes a horizontal frictional force on the spindle tending to accelerate precession. This causes the end *F* of the spindle to rise and push back the car against the wind-pressure. Thus the car turns over to meet the wind and is carried over beyond the vertical (perhaps considerably beyond) to a new inclined neutral position at which the moment due to gravity just balances the moment of the wind-pressure. This tilted position is reached, however, with a certain momentum which carries the car beyond it. Up to the instant of reaching this position the moment of the wind-pressure has been in excess of the opposite moment of the gravitational pull, and the impulse that has acted is measured by the angular displacement of the spindle along the guide. From the instant that the car swings past the new neutral position, the moment of the gravitative pull-over to the left exceeds that of the wind-pressure, and the car turning over to the left lifts the guide-plate G_1 off the roller R_1, and the gyrostat is left to itself, till a very slight further tilting over of the car brings a second curved guide-plate G_2 (fixed to the car) to bear on the lower side of a second roller R_2 fitting loosely on a non-rotating sleeve which is part of the frame *BAC*. This arrests the further turning over of the car and the gyrostat begins to precess back, but this time there is no force accelerating precession, so that the car remains tilted a little beyond the neutral position, till the roller R_2, in passing beyond the middle position, is carried clear of the end of the guide-plate G_2; the gyrostat is now left again to itself, but the car being released from the pressure of R_2 at once falls over a very little more to the left and thus brings to bear on the bottom of the rotating roller R_1 the third guide-plate G_3. The spindle now begins to roll over G_3; its precession is accelerated by friction, and the roller R_1 pushes back the guide-plate and so the car up to and beyond the new neutral position of equilibrium, when the turning moment on the car again begins to be reversed, and the car, being now pushed over by the wind to the right of this position, brings its fourth guide-plate G_4 to bear on the roller R_2, so that further turning is arrested, and the spindle *OF* precesses back to the middle position. In this way, the oscillations about the new neutral tilted position quickly diminish, and by thus arranging that the car shall overshoot the mark and then return in oscillations of diminishing amplitude, the time-integral of the upsetting couple is reduced to zero, which is the condition that the spindle shall be left at rest in the middle or "ready" position, after any adjustment. Each time the car has to move towards a new neutral position of equilibrium, work is done at the expense of the energy of spin of the disc; this, and the energy lost in frictional heat, is made up by the battery which maintains the rotation of the disc.

"It is evident that, instead of an upsetting couple due to wind-pressure, we may have one arising from a lateral shift of load on the car. The same process of self-adjustment will be gone through, the car inclining itself to the left if the shift of load is to the right; just as a man carrying a load on his right shoulder leans over to the left till the centre of gravity of his system is brought vertically over the supporting base formed by his feet.

"We have spoken so far of only a single gyrostat, but it is evident that if we endeavour to travel round a curve the spinning disc will maintain its plane of rotation unaltered in direction, and the car carrying the guides will sweep them round past the end of the spindle F, and thus the middle position will be lost. To remedy this Mr. Brennan employs a second similar gyrostat with an equal disc, spinning at an equal rate in the opposite direction, about a spindle which, when all is in equilibrium, is parallel to or even in a line with the spindle of the first disc. The pivoted frame BAC of the first is so linked to the corresponding frame of the second that any lateral tilt of the first is communicated to the second, but at the same time each of the discs is free to precess. The precession of the second disc is equal to that of the first, but in the opposite direction, and any deviation from this equality and opposition is prevented by toothed gearing which connects the axle ED of the first with the corresponding parallel axle of the second. Such a system offers no resistance to turning with the car on account of a curve in the track, while to any upsetting moment it behaves like a single gyrostat of double mass, and enables the car to meet the upsetting moment of the so-called centrifugal forces by leaning over towards the inner side of the curve, exactly as it leant over to meet a wind-pressure.

"It should, however, be observed that this adjustment does not get rid of the force tending to displace the rail laterally, and that this can only be completely met by sloping the track on which the rail is laid with exactly the same super-elevation as is required in an ordinary railroad curve (a slope which depends on the velocity prescribed). Mr. Brennan gets rid of the danger of upsetting, but not of the need of providing against displacement of the rail.

"It remains to examine what will happen when we pass from a model to a car of larger dimensions. Fortunately, the result works out very favourably, since we find that if we make the linear dimensions of everything n-times greater, we can afford to spin the gyrostats n-times slower and yet secure the same righting effect, with the same angular excursion and return of the spindle along the guides.

"This result is of great importance, for it means also that the centrifugal stresses in the real gyrostats need not be greater than in the model, and that the rate of spin may be reduced from 7000 per minute in the model to 875 per minute in a car of eight times the size. A greater rate in a smaller gyrostat is however a preferable option.

"In this explanatory outline we have been guided by the patent specifications already published; but we understand that Mr. Brennan has already made important improvements which will not be published till further protection has been obtained."

74. Note on the change of dimensions in a larger car compared with those of a model one. The efficiency of the erecting power of the gyrostat on the car, may be conveniently defined as the ratio of the impulse of the upsetting couple (due, say, to a gust of wind or a shift of load) to the resulting angular precessional displacement, relative to the car, of the axis of the gyrostat; for the larger the impulse required to produce a given displacement, the larger the resistance which the car offers to being upset.

Now the arrangement of the car is such, that after some time t the car has no angular momentum about the line of rail; hence the

$$\begin{aligned}\text{impulse of applied couple} &= \text{impulse of erecting couple}\\ &= \int I\omega\Omega\,dt\\ &= I\omega\int\Omega\,dt\\ &= I\omega\theta,\end{aligned}$$

Ω being the precessional velocity and θ the angle of precession.

Hence, the efficiency $= I\omega$.

If we now increase the linear dimensions n-times, the upsetting couple produced by a corresponding lateral displacement of load, say, will be n^4-times as great: for the displacement is n-times greater and the load is n^3-times greater.

Hence, for equal efficiency we must have $I\omega$ n^4-times greater. Now I is n^5-times greater; for $I = Mk^2$ where M is n^3-times and k is n-times as great. Consequently, if ω is n-times less, we shall have equal efficiency.

For further information concerning the Mono-railway, the reader is referred to Professor Perry's article in *Nature*, March 12, 1908.

CHAPTER VI.

STEADY MOTION OF A TOP.

75. The student is reminded at the commencement of this chapter that, when a solid body is under consideration,

(i) Angular velocity about any line means total angular velocity—not relative to some moving plane, unless this is expressly stated.

(ii) Angular velocity about a line which is moving means (total) angular velocity about the line fixed in space, with which the moving line happens to be coinciding at the instant in question.

76. In the preceding chapter the bodies whose rotation we have discussed have been symmetrical bodies, as, for example, a fly-wheel; and all the rotations have been about an axis of symmetry, *i.e.* the axle. If the axis were not an axis of symmetry, an angular *velocity* about this axis would in general involve angular *momentum* (about this axis, and also) about the two axes perpendicular to it, as is shown in the next article. In this chapter we propose to discuss the equations of motion of an ordinary spinning top, in which case it is clear that only the axle of the top is an axis of symmetry, and any other axis is not. But we shall see in Art. 79 that, since the top is a solid of revolution, any axis perpendicular to the axle of the top is the same for our purpose as if it were an axis of symmetry, and angular velocity about such an axis involves (of course angular momentum about that axis, but) *no angular momentum about perpendicular axes.*

77. Rotation about one axis involves in general angular momenta about other axes at right angles to the first. In general, if a body has at any instant an angular velocity about a given axis, this *velocity* involves an *angular momentum* about each of *two* lines perpendicular to the original axis and to each other.

For let OZ be the original axis of rotation about which the body has an angular velocity in the positive direction.

Let OX, OY, be two straight lines perpendicular to OZ and to each other.

Let x, y, z be the co-ordinates of any particle P of the body, referred to these axes.

Then the angular momentum of the particle P is:

About OZ, $\qquad m\omega r \,.\, r$ or $mr^2\omega$.

About OX,
$$m\omega r \cos\theta \,.\, z$$
$$= -m\omega r \,.\, \frac{x}{r} \,.\, z = -m\omega z \,.\, x.$$

About OY,
$$m\omega r \sin\theta \,.\, z$$
$$= -m\omega r \,.\, \frac{y}{r} \,.\, z = -m\omega z \,.\, y.$$

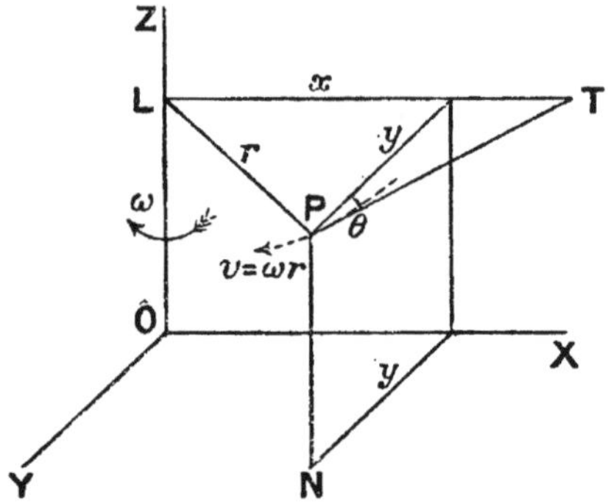

FIG. 37.

Hence it is seen that, owing to the rotation of the body about OZ, there is a contribution of angular momentum about OX equal to $-\Sigma m\omega z \,.\, x$, and about OY equal to $-\Sigma m\omega z \,.\, y$, in addition to $\Sigma mr^2\omega$, or $I_z\omega$, about OZ.

The expressions Σmzx, Σmzy are called *products of inertia*, but we shall be able to neglect these in dealing with solids of revolution, as will be seen in Art. 79.

78. It should be noticed that although the rotation ω of the body about OZ involves in general *angular momentum* about both OX and OY, yet it involves no component angular *velocity* of the body about OX and OY; for the resolute of ω about OX or OY is $\omega \cos 90°$, which is zero.

This can be further illustrated by considering two points A, B moving in parallel lines with the same velocity so that AB is perpendicular always to the lines of motion. If the velocity at any instant is v, though the angular velocity of B about A is zero, the angular momentum of B about A is mvc, where c is the distance AB.

79. Particular case of a solid of revolution. If the body is a solid of revolution whose axle is the axis of rotation OZ, as, for instance, an ordinary top spinning vertically on its toe, it is clear (Fig. 38) that for every particle P revolving about OZ in direction (XY) and possessing a definite linear velocity in a definite direction in space, there is a corresponding particle, on the opposite side of the axle OZ to P and equidistant from it, which is moving in the opposite direction in space with the same velocity.

As far as rotation about OZ is concerned, these two moving particles cause angular momentum about OZ.

But when we consider the angular momentum about OX, OY, we see that the angular momenta of these particles cancel each other.

So for all the particles of this solid of revolution.

Hence, if the body rotates about its axle OZ with velocity ω, it has no other angular momentum due to ω besides $I_z\omega$, and we have already seen that it has no angular *velocity* due to this rotation besides ω.

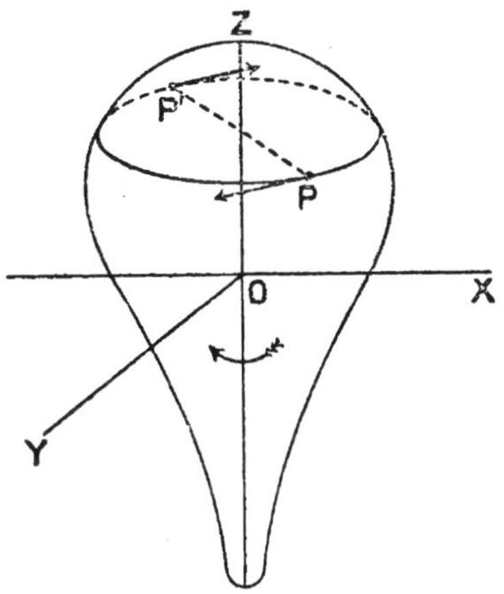

FIG. 38.

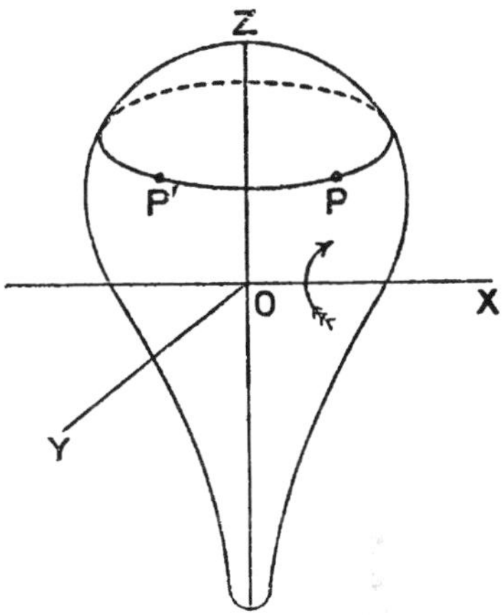

FIG. 39.

Now suppose the body has only a rotation about OX, but OZ is the axle of the solid of revolution (Fig. 39).

At a given instant any two corresponding particles P, P' are moving with equal and parallel velocities in planes parallel to YOZ.

Resolving these equal velocities in directions OY, OZ, we see that the angular momenta of the particles cancel about both OY and OZ.

Hence, in this case, the *whole* angular momentum of the solid due to the rotation ω is represented by $I_x\omega$. An axis such that rotation about it introduces no angular momentum about any perpendicular axis is called a *principal axis*.

It follows that in a top, the axle of the top is a principal axis, as is also any other axis perpendicular to it.

80. Moving axle. It is clear that if the *axle* of a rotating body moves in any manner it carries the body with it. In fact, the axle is equivalent to an indefinitely fine wire passing through the body.

81. Moving axis. It is sometimes convenient to resolve the angular velocity of a rotating body about three axes mutually at right angles, and these three axes form a frame of reference, which may itself perhaps be moving.

If we know the motion of the body relative to this frame, and the motion of the frame, we know the motion of the whole body.

If the frame of reference is moving, the axes are called *moving axes*, and the frame does not necessarily carry the body with it, for all the axes are not necessarily fixed in the body, but may be passing through it.

82. The velocities and momenta of a spinning top in steady motion. By the *steady motion* of a top is meant the motion in which the angular velocity ω about the axle of the top remains constant, while the axle has a constant inclination α to the vertical, and is carried round the vertical with constant angular velocity Ω (Fig. 40). This velocity Ω is called the azimuthal velocity, the vertical plane AGZ through the axle of the top being called the azimuthal plane. During steady motion, the centre of gravity G describes a horizontal circle. The lines GC, GA, of which GA is always moving through the material of the top but remaining with GC in the azimuthal plane, are taken as moving axes of reference.

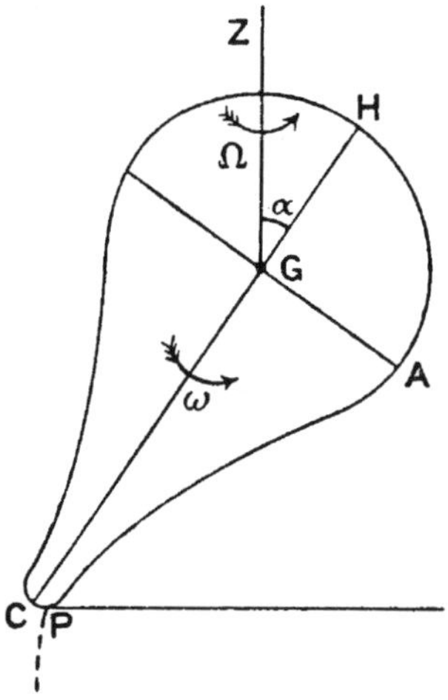

FIG. 40.

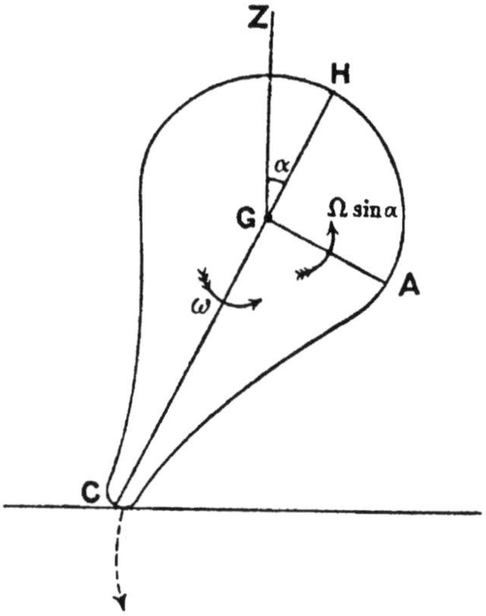

FIG. 41.

The velocities. The angular velocity Ω of the azimuthal plane about ZG (or $-\Omega$ about GZ, Art. 20), can be resolved (Fig. 41) into

$$(a)\ \Omega \sin \alpha \text{ about } GA,$$

$$(b)\ \Omega \cos \alpha \text{ about } GC.$$

If ω_a is the angular velocity of the top about GC *relative to the azimuthal plane*, then the total angular velocity of the top about GC is given by $\omega=\omega_a+\Omega\cos\alpha$, and the total angular velocity about GA is $\Omega\sin\alpha$.

Thus the angular velocity of the top is the resultant of

(i) the spin ω about the axle GC;

(ii) the spin $\Omega\sin\alpha$ about the axis GA.

About an axis perpendicular to GC and GA there is no angular velocity, since α is constant.*

The momenta. Let the moments of inertia about GC, GA be C, A, respectively.

Since GC, GA are principal axes, the angular velocity ω about GC gives an angular momentum component $C\omega$ about that axis (and none about any other axis); so the angular velocity $\Omega\sin\alpha$ about GA gives a component $A\Omega\sin\alpha$ about GA.

Thus the angular momentum of the top is the resultant of $C\omega$ and $A\Omega\sin\alpha$ about GC and GA respectively.

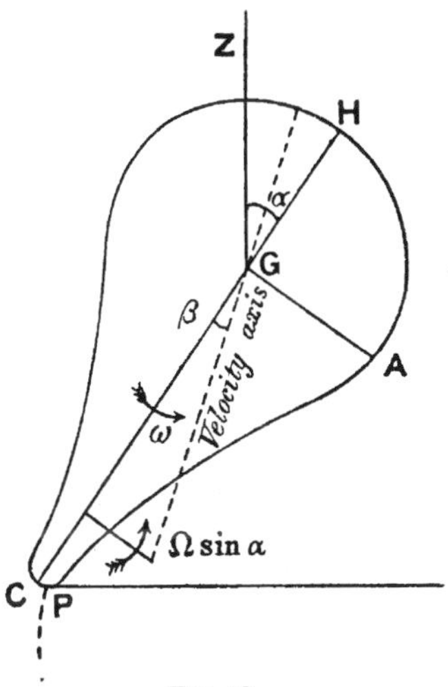

Fig. 42.

83. Velocity axis. The axis of component angular velocity lying in the azimuthal plane we shall call the *velocity axis.* It is clear (Fig. 42) that it makes an angle β with GC given by $\tan\beta=\dfrac{\Omega\sin\alpha}{\omega}$. The axis of *total* resultant angular velocity is called the *instantaneous axis* of rotation.

84. Momentum axis. The axis of component angular momen tum lying in the azimuthal plane we shall call the *momentum*

* The motion will perhaps be more easily realised if we consider a small roller such as is sometimes used for mounting photographs. In the accompanying figure, $Z'CH$, which corresponds to the azimuthal plane, is turning about $Z'C$ with velocity Ω, which has about GC a component velocity $\Omega\cos\alpha$; the roller corresponds to the top and turns about its axle GC with velocity ω_a *relative to the frame*, but with *total* velocity ω, where $\omega=\omega_a+\Omega\cos\alpha$. When we watch a top spinning, we do not think of its spin relative to the azimuthal plane, but of its total spin about its axle.

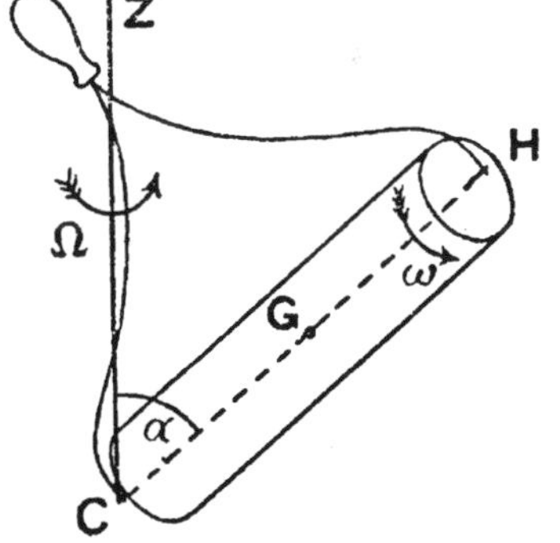

axis. It is clear (Fig. 43) that it makes an angle γ with GC given by $\tan\gamma = \dfrac{A\Omega\sin\alpha}{C\omega}$.

It will be seen that precession of the momentum axis necessitates a corresponding precession of the axle of the top, since both lie in the azimuthal plane. Hence we may discuss the motion of the top by considering the rotation of the momentum axis (see Art. 85). In steady motion the momentum axis coincides with the axis of resultant angular momentum.

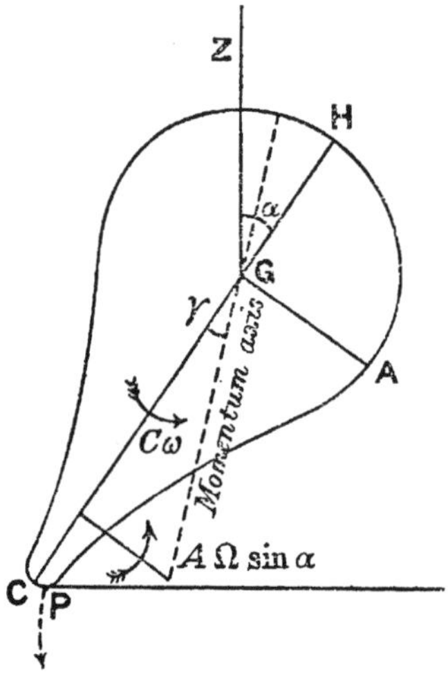

FIG. 43.

85. Torque required to rotate the momentum axis. Referring to Fig. 20, p. 37, let OX, OY, OZ be any three mutually perpendicular lines, and let OX be the momentum axis at any instant, and take OA to represent the magnitude (M) of the momentum.

Let Ω be the angular velocity of OX round OZ (see Art. 40), so that after time δt, the angular momentum is represented by OA', where the angle $AOA' = \Omega\delta t$.

Then if $A'B'$ be drawn parallel to AO, OB' represents the

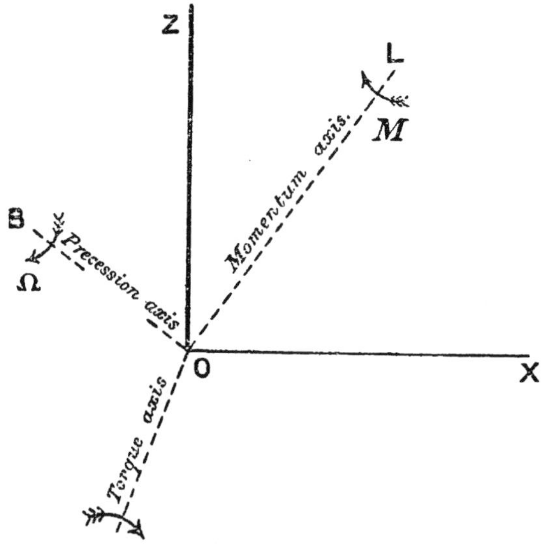

FIG. 44.

change of angular momentum in time δt, since OA and OB' are equivalent to OA'.

And, since $OB' = AA' = M\Omega\delta t$, the change of angular momentum in time δt is $M\Omega\delta t$ about OY.

$\therefore$ rate of change of angular momentum is $M\Omega$ about OY.

This requires for its production a torque $M\Omega$ about OY.

Hence (Fig. 44) if OL represents the momentum axis of a top whose toe is the fixed point O, the steady motion may be discussed by considering that OL is rotated about OB, the perpendicular axis in the azimuthal plane, by the torque about OY perpendicular to the azimuthal plane. It will be noticed that the gyroscopic resistance of the top to being turned about OY is $M\Omega$ (see Art. 45).

Cor. If the axes of resultant angular momentum and angular velocity be any two lines OP, OQ, the angular momentum and velocity can each be resolved into components about OX, OY, OZ and the torques producing the rotations of the component momenta about the three axes can be separately found as above, and equated to the corresponding components of the external acting torques.

This method of determining the motion of a top is the one most generally employed, since it is capable of easy extension to the case where the motion is not steady.

86. The steady motion of a top spinning on a fine fixed point. Consider the steady motion of a top spinning with its axle inclined at an angle α to the vertical on a fine point O which is

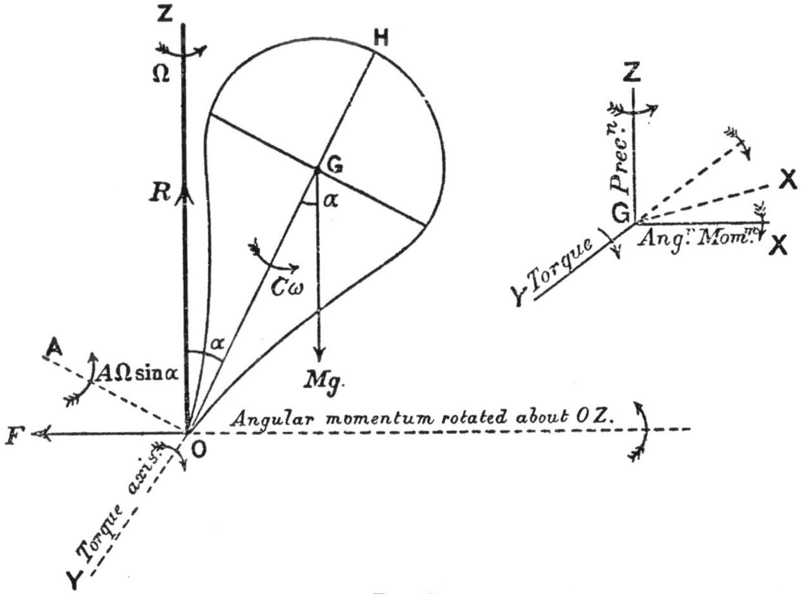

FIG. 45.

fixed. An instance of such a motion will be that of a top spinning on a table which is rough enough to prevent slipping, while the toe is considered too fine for the friction to have a moment about the axle, and therefore diminish the spin (see Art. 128). Let the directions of rotation be those marked in Fig. 45.

The external forces acting on the top are:

(i) its weight Mg at G;

(ii) a vertical reaction R at O;

(iii) a horizontal reaction F at O.

This last must act in the direction marked, since G describes a horizontal circle about OZ, and F, the force causing it to do so, must be parallel to the inward radius.

From Art. 82 we see that the momenta are $C\omega$ about GO, and $A\Omega \sin \alpha$ about AO, where A is the moment of inertia about the axis through O.

These give component momenta $A\Omega \sin \alpha \cos \alpha - C\omega \sin \alpha$ about OX, and a component about OZ, the latter of which need not be considered, as it is not rotated and no torque therefore is required on account of it.

The former is rotated with angular velocity $-\Omega$ about OZ, and this requires a torque

$$(A\Omega \sin \alpha \cos \alpha - C\omega \sin \alpha)(-\Omega)$$

about OY (Art. 37).

But this torque is $Mga \sin \alpha$, where $OG = a$.

Hence, $\quad C\omega\Omega \sin \alpha - A\Omega^2 \sin \alpha \cos \alpha = Mga \sin \alpha,$

whence

$$\text{either} \quad \sin \alpha = 0,$$

$$\text{or} \quad C\omega\Omega - A\Omega^2 \cos \alpha = Mga.$$

The former gives $\alpha = 0$ or π, both of which are possible angles for steady motion.

The latter gives a quadratic for Ω, *i.e.*

$$A \cos \alpha . \Omega^2 - C\omega . \Omega + Mag = 0,$$

showing that there are, in genèral, two possible angular velocities of precession, *i.e.*

$$\Omega = \frac{C\omega \pm \sqrt{C^2\omega^2 - 4A \cos \alpha . Mag}}{2A \cos \alpha}.$$

If $\cos \alpha$ is negative, *i.e.* G below O, Ω is always real.

If $\cos \alpha$ is positive, then for real values of Ω we must have

$$C^2\omega^2 > 4A \cos \alpha . Mag,$$

$$\text{or} \quad \omega > \frac{2\sqrt{AMag . \cos \alpha}}{C}.$$

If ω has a smaller value than this, the top cannot spin steadily (compare Art. 102).

The larger or the smaller value of Ω will be maintained according to the particular initial conditions of motion. Thus if the top is started at an angle a with angular velocity ω, and either of the above values of Ω, it will continue to spin steadily.

From the above quadratic equation we see that during steady motion

$$\cos a = \frac{C\omega\Omega - Mag}{A\Omega^2},$$

showing that if C is increased a diminishes,
and if A „ a increases;

hence it appears that a top with a longer leg spins at a greater angle to the vertical.

87. When the majority of rotations in a problem under consideration are right-handed it will be found convenient to work throughout with right-handed axes. Whichever rotation is employed, the precessional velocity must be considered negative if it does not turn the angular momentum rotated, towards the torque axis both being drawn in the same sense.

88. Equation deduced from gyroscopic resistance. The equation of the preceding article may be obtained by considering that the torque $Mga \sin a$ is tending to turn about its axis the two (right-handed) components of angular momentum $C\omega$, and $A\Omega \sin a$; but these components, instead of being turned about OY, are precessing about OA and OG with velocities $\Omega \sin a$ and $\Omega \cos a$ respectively.

Hence, the gyroscopic resistances to being turned are respectively $C\omega\Omega \sin a$ and $A\Omega^2 \sin a \cos a$ (Art. 45).

If the torque axis is drawn right-handed we see that OG sets itself *towards* the torque, while OA sets itself *away from* the torque.

Thus, $C\omega\Omega \sin a$ must be considered a positive resistance,
$A\Omega^2 \sin a \cos a$ must be considered a negative resistance.

Since there is no change in angular momentum about the torque axis we have

$$Mga \sin a - C\omega\Omega \sin a + A\Omega^2 \sin a \cos a = 0,$$

$$\text{or} \quad Mga = C\omega\Omega - A\Omega^2 \cos a, \quad \text{as before.}$$

The general equations of which the above are a particular case are given in Art. 125.

89. The equation of Art. 86 might also be obtained by rotating the momentum axis instead of the horizontal component of angular momentum.

In Fig. 46, let AO be at any instant the axis of precession *

* It should be noticed that the term "axis of precession," which is suggested by astronomy, is not strictly in accordance with astronomical usage, where the axis of the earth is said to precess about a line perpendicular to the plane of its orbit, and not to its own axis. In the same way a top spinning at an inclination to the vertical can be said to precess about the vertical; but we are taking the axis of precession to be the axis perpendicular to the rotating torque and the angular momentum rotated.

round which the angular momentum $C\omega \sec\gamma$ is rotated with velocity $\Omega \sin(\alpha-\gamma)$.

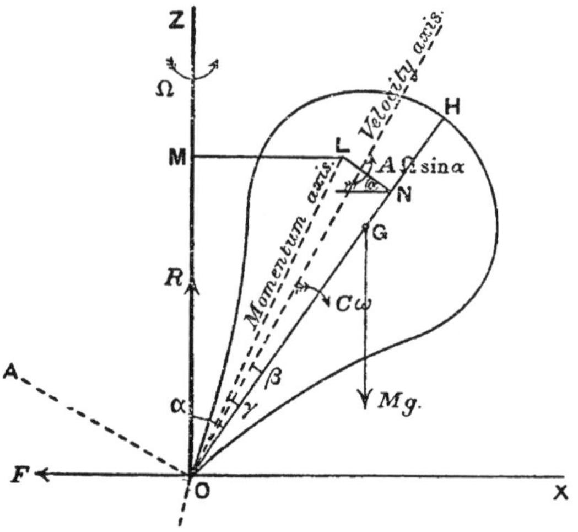

FIG. 46.

We have, employing right-handed rotations,

$$C\omega \sec\gamma \,.\, \Omega \sin(\alpha-\gamma) = Mga \sin\alpha,$$

$$C\omega\Omega(\sin\alpha - \cos\alpha \,.\, \tan\gamma) = Mga \sin\alpha,$$

$$\text{or} \quad C\omega\Omega\left(\sin\alpha - \cos\alpha \frac{A\Omega \sin\alpha}{C\omega}\right) = Mga \sin\alpha,$$

$$\textit{i.e.} \quad C\omega\Omega - A\Omega^2 \cos\alpha = Mga, \quad \text{as before.}$$

It will be noticed that the first of the above equations expresses the fact that the gyroscopic resistance is equal to the external torque.

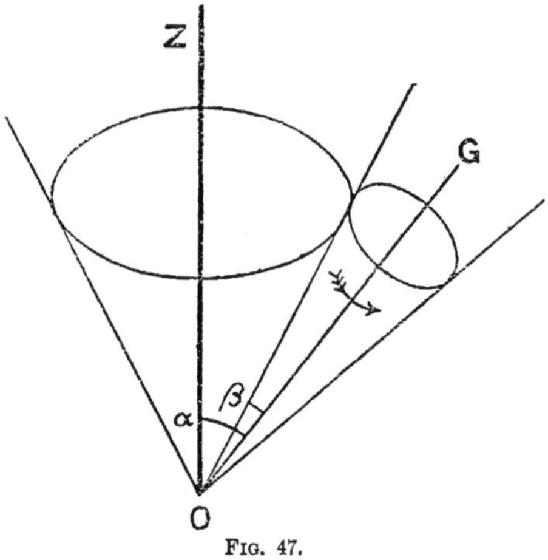

FIG. 47.

90. We see (Fig. 46) that the velocity axis is in this case the instantaneous axis, and makes a constant angle $(\alpha-\beta)$ with the vertical OZ: it therefore describes a cone *fixed in space* having OZ as axis and vertical angle $2(\alpha-\beta)$.

It is also inclined to OG at a constant angle β, and thus describes a cone *fixed in the body* having OG as axis and vertical angle 2β.

Since points of the body on the velocity axis are instantaneously at rest, the motion can be represented by the rolling of the latter cone on the former (Fig. 47).

[**24.**] Does all this hold as regards the momentum axis?

[**25.**] With what angular velocity does the moving cone roll round the fixed one?

91. Analogy of the hodograph. If a particle P (Fig. 48) is moving along a curve $P_1P_2P_3\ldots$ (not necessarily in one plane), and if from a point O fixed in space straight lines OA_1, OA_2, OA_3 are drawn to represent the velocities at the points

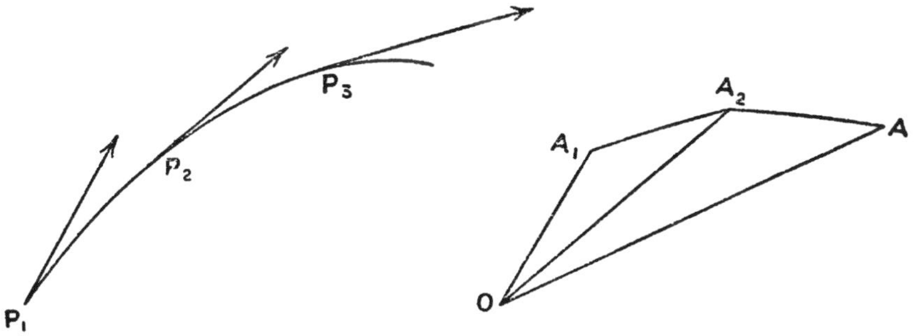

FIG. 48.

$P_1P_2P_3\ldots$, then the *velocity* of the point A along the curve $A_1A_2A_3\ldots$ represents the *acceleration* of the particle P as it moves along the curve $P_1P_2P_3,\ldots$, since it represents the rate of change of the velocity. The path $A_1A_2A_3\ldots$ is called the *hodograph* of P.

Similarly, if OA_1, OA_2, represent the *momentum* of the particle, then the velocity of the point A along its path represents the *force* acting on the particle P since it represents rate of change of momentum. Thus if OA_1, $OA_2,\ldots$, be successive positions of the axis of resultant angular momentum, and $OA_1,\ldots$ represent the angular momentum of a body at successive instants, then the velocity of A in its path represents the torque acting on the body, since it represents the rate of change of angular momentum.

92. We will now apply this analogy of the hodograph to obtain the equation of Art. 86.

In Fig. 46 let OL, which coincides with the momentum axis (Art. 84), represent at any instant the resultant angular momentum of the top. Draw LM and LN perpendicular to OZ and OG respectively. Then OM, ML are the vertical and horizontal components of the total angular momentum, and ON, NL are the components along and perpendicular to OH. The linear velocity of L is $LM.\Omega$, and represents (Art. 91) the torque causing precession, *i.e.* $Mga \sin \alpha$.

Now $ML =$ sum of horizontal projections of ON and NL

$$-ON \sin a - NL \cos a$$

$$-C\omega \sin a - A\Omega \sin a \cos a \text{ (Art. 82).}$$

Hence $$(C\omega \sin a - A\Omega \sin a \cos a)\Omega = Mga \sin a,$$

or $$C\omega\Omega - A\Omega^2 \cos a = Mga.$$

93. Steady motion of a solid of revolution spinning on a rough horizontal plane. In Fig. 49 let GC be the axle of the solid, a the angle it makes with vertical, P the point of contact at any moment with the plane; let $GE = c$, PL perpendicular to $AG = y$, and C and A be the principal moments of inertia through G.

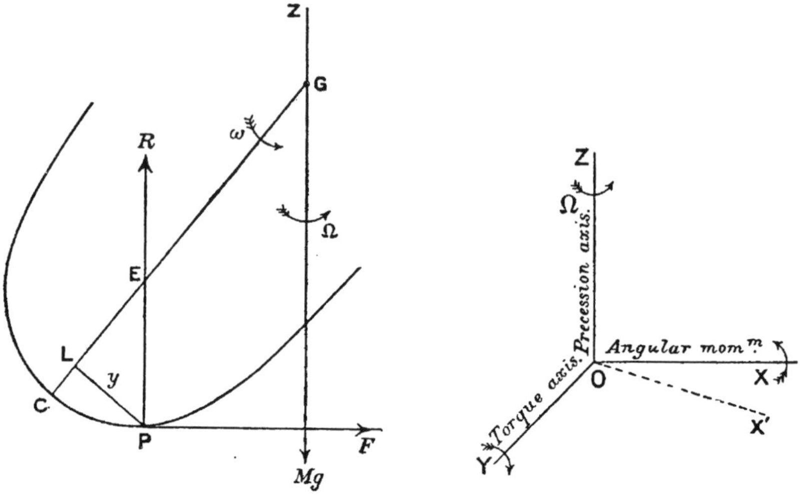

FIG. 49.

Here, since there is steady motion, ω is constant, and therefore there is no component of F *perpendicular* to the azimuthal plane, the reaction R being along the normal PE.

Since G describes a horizontal circle uniformly, if r is its radius, we have $$F = M\Omega^2 r.$$

Also, since G does not move vertically,

$$R = Mg.$$

Considering the angular momentum about GX the horizontal through G, as being made to rotate with velocity Ω about the vertical through G, we have (rotations right-handed) ·

angular momentum rotated

$$= C\omega \sin a - A\Omega \sin a \cos a \text{ about } GX,$$

velocity of rotation $= \Omega$ about GZ,

torque producing rotation

$$= Rc \sin a - F(c \cos a + y \operatorname{cosec} a) \text{ about } YG.$$

The other components of angular momentum are not rotated. Hence, paying due consideration to sign,

$$(C\omega \sin a - A\Omega \sin a \cos a)\Omega = Rc \sin a - F(c \cos a + y \operatorname{cosec} a),$$

or

$$C\omega\Omega \sin a - A\Omega^2 \sin a \cos a = Mgc \sin a - M\Omega^2 r(c \cos a + y \operatorname{cosec} a).$$

We have, in addition, since P is for the instant at rest, the geometrical condition, that

the velocity of P relative to L + that of L relative to G

+ that of G relative to the centre of G's circle = 0,

or
$$y\omega - \Omega \sin a(c + y \,.\, \cot a) - \Omega r = 0.$$

94. Steady motion on a surface of revolution. It should be noticed that steady motion can be maintained on any rough surface of revolution in exactly the same way as on a rough horizontal plane. For (Fig. 50) when the steady motion has once started, friction, as before, will act only in the azimuthal plane, the forces and couples acting during successive positions of the top will be the same, and the precessional motion will remain unaltered.

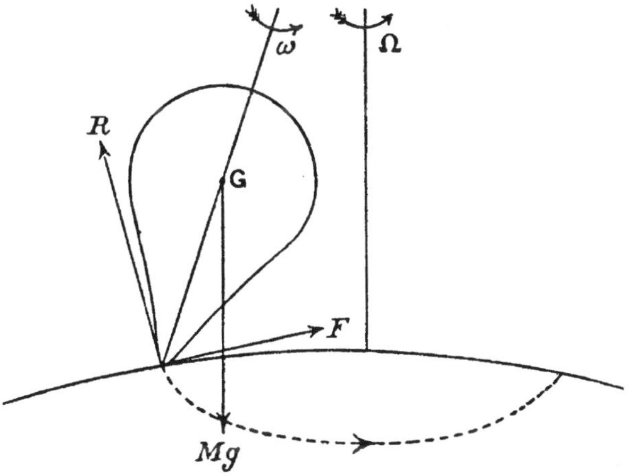

FIG. 50.

[26.] If the surface of revolution be smooth, under what conditions (if any) will steady motion be possible: (i) if the surface is convex to the top; (ii) concave? What direction does the precession take in the latter case?

95. The motion of a system of two or more gyrostats. *Four equal gyrostats have for axes the sides of a light rhombus ODEF, formed of rods jointed together, which hang from O, and all four are set spinning with equal angular velocities ω, and in such a way that all would be spinning in the same direction* *if the angles at O and E were zero. The mass of each gyrostat is M, and a mass M′ is suspended from E.*

Prove that if the angles at O and E are each 2α the whole can move with a steady precession Ω provided that

$$(A+Ma^2)\Omega^2\cos\alpha+C\omega\Omega=(2M+M')ag,$$

2a being the length of a side and C and A the principal moments of inertia of each gyrostat.

Let us consider the gyrostats spinning in the direction indicated in Fig. 51. Then (Fig. 51 a) taking the angular momentum rotated and the rotating torque right-handed about OX and YO respectively, it follows that Ω is about OZ.

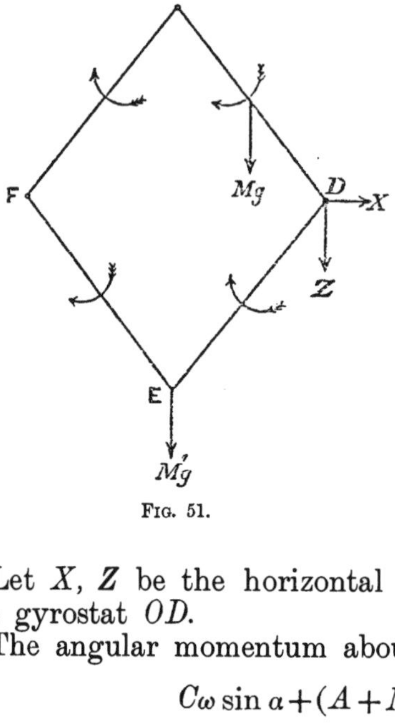

FIG. 51.

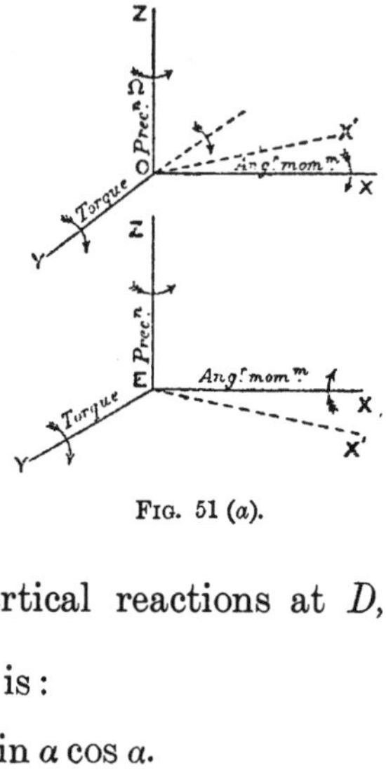

FIG. 51 (a).

Let X, Z be the horizontal and vertical reactions at D, on the gyrostat OD.

The angular momentum about OX is:

$$C\omega\sin\alpha+(A+Ma^2)\Omega\sin\alpha\cos\alpha.$$

The torque about YO is:

$$Mga\sin\alpha+Z2a\sin\alpha-X2a\cos\alpha;$$

$$\cdot\ [C\omega\sin\alpha+(A+Ma^2)\Omega\sin\alpha\cos\alpha]\Omega$$
$$=Mga\sin\alpha+Z2a\sin\alpha-X2a\cos\alpha.\ \ldots\ldots\text{(i)}$$

Now consider the lower gyrostat ED and the torque turning it about E.

Referring to Fig. 51 (a) for signs and using left-handed rotations, Ω is negative. The angular momentum about EX

$$=C\omega\sin\alpha+(A+Ma^2)\Omega\sin\alpha\cos\alpha.$$

Moment of torque about EY

$$=Mga\sin\alpha-Z2a\sin\alpha-X2a\cos\alpha;$$

$$\therefore\ [C\omega\sin\alpha+(A+Ma^2)\Omega\sin\alpha\cos\alpha](-\Omega)$$
$$=Mga\sin\alpha-Z2a\sin\alpha-X2a\cos\alpha.\ldots\ldots(ii)$$

Subtracting (ii) from (i),

$$2[C\omega\sin\alpha+(A+Ma^2)\Omega\sin\alpha\cos\alpha]\Omega=4aZ\sin\alpha.$$

But the two lower gyrostats and $M'g$ are supported by the two vertical reactions at F and D.

$$\therefore\ 2Z=2Mg+M'g.$$

Hence the condition for steady motion becomes

$$C\omega\Omega+(A+Ma^2)\Omega^2\cos\alpha=(2M+M')ag,$$

since α is not zero.

[27.] What are the reactions on the axes of the gyrostats at right angles to the azimuthal plane?

[28.] What would be the effect of the two right-hand ones spinning one way and the other two the opposite way?

96. Cone rolling on a rough horizontal plane. *A homogeneous right circular cone rolls on a perfectly rough horizontal plane. Prove that if the vertex remain in contact with the plane*

$$T^2>\frac{16\pi^2k^2\cos\alpha}{ag(4-3\cos^2\alpha)},$$

where T is the time of a complete revolution of the axis, a the radius of the base, α the semi-vertical angle of the cone, and k the radius of gyration about a generating line.

Consider the steady motion when the vertex O is *just not* touching the plane (Fig. 52).

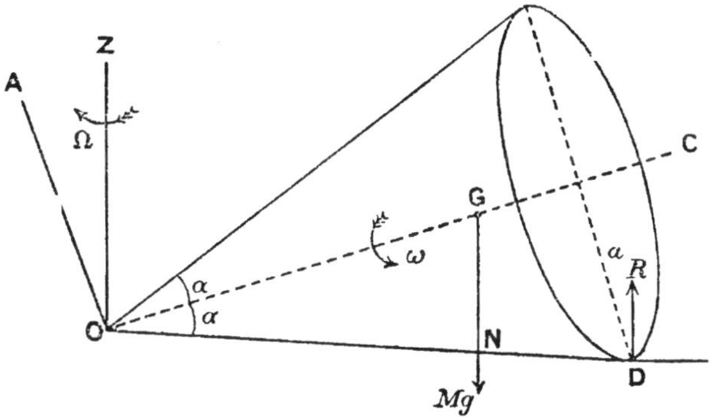

FIG. 52.

If ω is the spin about OC and Ω the azimuthal velocity, the angular momenta are $C\omega$ and $-A\Omega\cos\alpha$ about OC and OA, where C, A, are the principal moments of inertia at O.

The component momenta are therefore

$$C\omega \cos\alpha + A\Omega \cos\alpha \sin\alpha \quad \text{and} \quad C\omega \sin\alpha + A\Omega \cos^2\alpha$$

about OX, OZ respectively.

The latter is not rotated, but the former is rotated with angular velocity Ω about OZ and this (Art. 37) requires a torque $\Omega\,[C\omega \cos\alpha + A\Omega \cos\alpha \sin\alpha]$ about OY.

But since O is just not in contact with the plane, the pressure of the plane acts through D, and the moment of the acting torque is

$$Mg\,DN = Mg\left[\frac{a}{\sin\alpha} - OG\cos\alpha\right]$$

$$= Mg\left[\frac{a}{\sin\alpha} - \frac{3}{4}a\cot\alpha\cos\alpha\right]$$

$$= \frac{Mga}{4\sin\alpha}(4 - 3\cos^2\alpha).$$

$$\therefore\ \Omega\,[C\omega\cos\alpha + A\Omega\cos\alpha\sin\alpha] = \frac{Mga}{4\sin\alpha}(4 - 3\cos^2\alpha). \qquad \text{............(i)}$$

Now consider the motion of the point D of the cone. Since the cone has rotations ω about OC and $-\Omega\cos\alpha$ about OA, the parts of the velocity of D due to each, parallel to OY, are respectively $-a\omega$ and $+(a\cot\alpha)(\Omega\cos\alpha)$.

Hence, since D is momentarily at rest,

$$-a\omega + a\Omega\cot\alpha\cos\alpha = 0,$$

whence

$$\omega = +\frac{\Omega\cos^2\alpha}{\sin\alpha}.$$

$\therefore$ substituting in (i),

$$\Omega^2\left[\frac{\cos^2\alpha}{\sin\alpha}\ C\cos\alpha + A\sin\alpha\cos\alpha\right] = \frac{Mga}{4\sin\alpha}(4 - 3\cos^2\alpha),$$

and since

$$A\sin^2\alpha + C\cos^2\alpha = \text{moment of inertia about } OD$$
$$= Mk^2,$$

$$\Omega^2 k^2\cos\alpha = \frac{ga}{4}(4 - 3\cos^2\alpha),$$

and the periodic time T is $2\pi/\Omega$,

$$\therefore\ T^2 = 4\pi^2\frac{4k^2\cos\alpha}{ga(4 - 3\cos^2\alpha)}$$

$$= \frac{16\pi^2k^2\cos\alpha}{ga(4 - 3\cos^2\alpha)},$$

and when the cone is *on* the plane, Ω being smaller, T^2 is greater than this value.

[29.] When the cone touches the plane what torque causes it to precess?

EXAMPLES.

NOTE. *In the following examples ω and Ω are taken so that they would be in the same direction if their axes coincided.*

1. Employ the preceding methods to show that the condition for steady motion of a top with a blunt peg (or any solid of revolution) spinning with velocity ω on a smooth horizontal plane is $C\omega\Omega - A\Omega^2 \cos\alpha = Mgc$, where C and A are moments of inertia about the axes in the azimuthal plane through the centre of gravity G, c is the distance between G and the point of intersection of the axle and the normal reaction, and α is the angle the axle makes with the vertical. Explain the similarity to the equation in Art. 86.

NOTE. Since the plane is smooth the top "skids" round during the whole motion; there are no frictional forces at the point of contact, and therefore the centre of gravity G remains at rest—the axle of the top describing a cone, vertex G.

2. A thin circular disc, radius c, spins with velocity ω about an axis through its centre perpendicular to its plane, while the rim is in contact with a smooth horizontal plane. If Ω is the azimuthal velocity, show by the above methods that the condition for steady motion is

$$c\Omega^2 \sin\alpha - 2c\omega\Omega = 4g \tan\alpha,$$

where α is the angle the plane of the disc makes with the vertical.

3. A circular wire ring of radius a rolls on a rough horizontal plane, so that its plane maintains a constant inclination α to the vertical; if ω be the angular velocity of the ring, and Ω the azimuthal motion of its centre, prove that

$$4a\omega\Omega \cos\alpha - a\Omega^2 \sin\alpha \cos\alpha = 2g \sin\alpha.$$

4. Show that the condition for steady motion of a disc spinning as in example 2, but on a rough horizontal plane, is

$$c^2\omega^2 \cot\alpha (6r + c \sin\alpha) = 4gr^2,$$

where r is the radius of the circle described by the centre of the disc.

5. Obtain the equation for steady motion, as in Art. 95, for the case of six gyrostats.

Would it be possible with steady motion to spin (a) the two vertical gyrostats in opposite directions? (b) the two upper ones in the opposite direction to the two lower ones?

6. One gyrostat, axis $2a$, mass M_1, spins on a fine fixed point, while another, axis $2b$, mass M_2, spins on the head of the former. Supposing a steady azimuthal motion Ω to be maintained when each makes an angle α with the vertical, determine equations for ω_1, ω_2, the spins of the two gyrostats, C_1, A_1, etc. being the principal moments of inertia through the centres of gravity.

Discuss the following points:

(i) If the upper one spun the reverse way to the lower one, would steady motion be possible?

(ii) Would steady motion be possible if the upper one made an angle β with the vertical?

(iii) If the two gyrostats were alike and both inclined at an angle α to the vertical, would $\omega_1 = \omega_2$?

7. A gyrostat has a universal joint at a point in its axis and the joint is attached by a string of length l to a fixed point; an angular velocity ω is given to the body and Ω is the angular velocity of the centre of gravity about the vertical, θ the angle which the string, ϕ the angle which the axis of the body, makes with the vertical, a the distance between the centre of gravity and the point of suspension, C being the moment of inertia about the axis of figure. Show that when motion is steady,

$$\Omega \sin\phi(C\omega - A\Omega\cos\phi)\cos\theta = Mga\sin(\phi-\theta),$$
$$\Omega^2(l\sin\theta + a\sin\phi) = g\tan\theta.$$

If another gyrostat were spun on the head of this one, which way would the spin have to be in order to maintain steady motion?

8. A heavy sphere is held in contact with a rough circular wire which is fixed in a horizontal plane, and a horizontal impulse is then applied to the sphere, causing it to roll round steadily. If c is the radius of the ring and b that of the sphere, and if α is the constant inclination to the vertical of the radius through the point of contact, Ω the angular velocity of the point of contact, show that the magnitude of the impulse is such as would impart to the sphere, if it were free, a velocity

$$\tfrac{7}{5}\Omega(c - b\sin\alpha),$$

and that the relation between Ω and α is given by

$$7\Omega^2(c - b\sin\alpha) = 5g\tan\alpha.$$

NOTE. The point of contact of the sphere with the wire is (instantaneously) a fixed point.

9. A homogeneous sphere of radius a is loaded at a point on its surface by a particle whose mass is $\frac{1}{n}$th of its own; if it move steadily on a smooth horizontal plane, the diameter through the particle making a constant angle α with the vertical, and the sphere rotating about it with uniform angular velocity ω, prove that ω^2 must be not less than

$$5g\cos\alpha(2n+7)/an(n+1)$$

if the particle is at the upper extremity of the diameter; and show that the particle will revolve round the vertical in one or other of two periods whose sum is $4\pi na\omega/5g$.

NOTE. The moment of inertia for the whole body about an axis through its centre of gravity G, perpendicular to the diameter through G, is $\frac{M}{n+1}\cdot\frac{2n+7}{5}a^2$, where M is the mass of the sphere.

10. A sphere is rotating within a spherical concentric light shell of radius a, placed on a rough horizontal plane, about an axis through the common centre. Show that if the centre of the sphere describes a circle of radius r with uniform velocity v, while spinning with velocity ω, then the inclination of the axis of rotation is given by

$$k^2\sin\alpha(v\cos\alpha - r\omega) = var,$$

where k is the radius of gyration of the sphere.

11. A homogeneous right circular cone, radius of base a, height h, spinning about its axis with velocity ω, maintains a constant azimuthal motion Ω with its axis inclined to the vertical at an angle β and its base in contact with a rough horizontal plane. Show that the necessary condition is

$$\frac{3}{10}a^2\omega\Omega\sin\beta - \frac{3}{20}\left\{a^2+\frac{h^2}{4}\right\}\Omega^2\sin\beta\cos\beta$$
$$= g\left(a\cos\beta \backsim \frac{h}{4}\sin\beta\right) - \Omega^2 . r\left(\frac{h}{4}\cos\beta + a\sin\beta\right),$$

r being the radius of the circle described by the centre of gravity of the cone.

Additional problems on steady motion will be found among the Miscellaneous Examples at the end of Chapter IX.

97. General motion of a top spinning on a fine fixed point. Hitherto we have only discussed the motion of a top when its axle makes a constant angle α with the vertical. We will now consider the case where this angle may be a variable quantity θ throughout the motion.

It should be particularly remembered throughout this chapter that *all* frictions are neglected except where otherwise stated.

Let θ be the angle which the axle makes with the vertical through O (Fig. 53), ω the spin of the top, and $\dot{\psi}$ the azimuthal

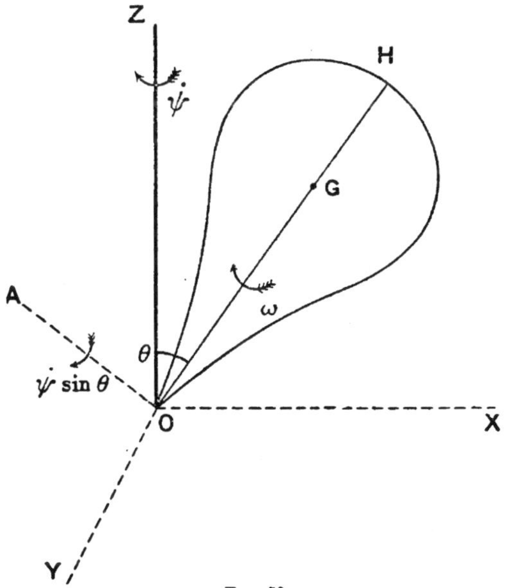

FIG. 53.

velocity, ψ being the angle the azimuthal plane makes with some plane fixed in space and passing through OZ.

Since there are no forces acting about the axle, with regard to which the top is symmetrical, ω will remain constant.*

* A more rigorous proof of this is given in Art. 128, where it is shown that the condition of symmetry is essential.

The reader is reminded that, as in Art. 82, ω is equal to $\omega_a + \dot{\psi}\cos\theta$, where ω_a is the (now variable) velocity relative to the azimuthal plane.

The component velocities of the top are:

$$\omega \text{ about } OH \cdot$$
$$\dot{\psi} \sin\theta \text{ about } OA;$$
$$\dot{\theta} \text{ about } OY;$$

where OY is perpendicular to the azimuthal plane.

The component angular momenta are:

$$C\omega \text{ about } OH;$$
$$A\dot{\psi}\sin\theta \text{ about } OA;$$
$$A\dot{\theta} \text{ about } OY.$$

Since there are only two unknown quantities (θ, $\dot{\psi}$), only two equations are necessary. Hence the whole motion of the axle of the top may be determined by the two following considerations ·

(α) The angular momentum about OZ remains constant.

(β) The sum of the kinetic and potential energies of the top is constant.

98. Motion of the top when the centre of gravity G has no initial velocity. In this case we see from (α) that

$$C\omega\cos\theta + A\dot{\psi}\sin^2\theta = C\omega\cos\theta_0, \quad \text{.................(i)}$$

if θ_0 is the inclination to the vertical at which the top is first spun, the axle of the top being initially at rest.

From (β) we have

$$\tfrac{1}{2}C\omega^2 + \tfrac{1}{2}A\dot{\psi}^2\sin^2\theta + \tfrac{1}{2}A\dot{\theta}^2 + Mga\cos\theta = \tfrac{1}{2}C\omega^2 + Mga\cos\theta_0,$$

or
$$A\dot{\psi}^2\sin^2\theta + A\dot{\theta}^2 = 2Mga(\cos\theta_0 - \cos\theta). \quad \text{...........(ii)}$$

Equation (i) determines the azimuthal velocity in terms of θ,

$$\text{i.e.}\quad \dot{\psi} = \frac{C\omega(\cos\theta_0 - \cos\theta)}{A\sin^2\theta}, \quad \text{....................(iii)}$$

and the above equations determine the complete motion of the top.

99. The top will oscillate between two limiting values of θ The elimination of $\dot{\psi}$ between (i) and (ii) yields

$$C^2\omega^2(\cos\theta_0 - \cos\theta)^2 + A^2\dot{\theta}^2\sin^2\theta = 2MAga(\cos\theta_0 - \cos\theta)\sin^2\theta.$$

The stationary values of θ are given by $\dot{\theta} = 0$,

i.e. by $(\cos\theta - \cos\theta_0)[2MAga\sin^2\theta - C^2\omega^2(\cos\theta_0 - \cos\theta)] = 0$,

$$\text{i.e.}\quad \theta = \theta_0 \text{ or } \sin^2\theta = 2\lambda(\cos\theta_0 - \cos\theta),$$

where 2λ is written for $C^2\omega^2/2MAga$.

The latter is $\cos^2\theta - 2\lambda\cos\theta + 2\lambda\cos\theta_0 - 1 = 0$,

whence
$$\cos\theta = \lambda \pm \sqrt{1 - 2\lambda\cos\theta_0 + \lambda^2}.$$

Now,

$$\lambda+\sqrt{1-2\lambda\cos\theta_0+\lambda^2}>\lambda+\sqrt{1-2\lambda+\lambda^2} \text{ since } \cos\theta_0<1,$$

$$\text{i.e. } >\lambda+1-\lambda,$$

$$\text{i.e. } >1.$$

Hence the former root, though giving a real value of $\cos\theta$, gives an imaginary value of θ.

The latter root gives a real value θ_1 for θ where

$$\cos\theta_1=\lambda-\sqrt{1-2\lambda\cos\theta_0+\lambda^2}.$$

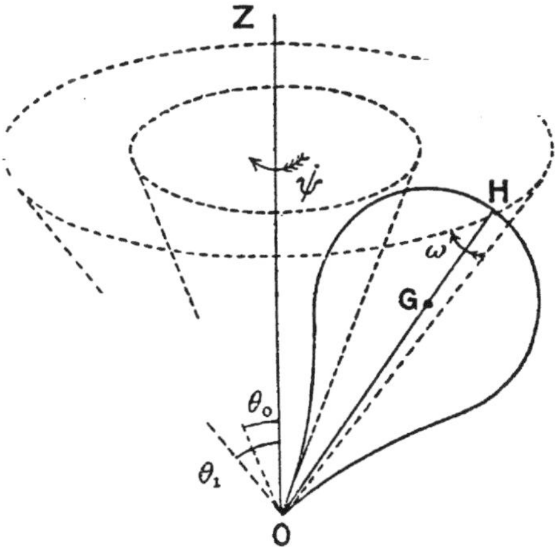

FIG. 54.

Hence, $\dot{\theta}=0$ (*i.e.* the axle momentarily ceases to rise or fall) when $\theta=\theta_0$ (upper limit) or $\theta=\theta_1$ (lower limit): that is, the axle of the top oscillates in the azimuthal plane between two cones (Fig. 54) whose semi-vertical angles are θ_0 and θ_1, while the azimuthal plane itself rotates with varying velocity $\dot{\psi}$.

The lower limit is above, at, or below the horizontal position according as $2\lambda\cos\theta_0>$, $=$, or <1, this being the condition that $\cos\theta_1$ is positive, zero, or negative.

100. To determine the value of $\dot{\psi}$ at the limiting positions.

We have the equations

$$A\dot{\psi}\sin^2\theta=C\omega(\cos\theta_0-\cos\theta), \quad\text{.................. (i)}$$

$$A\dot{\psi}^2\sin^2\theta+A\dot{\theta}^2=2Mga(\cos\theta_0-\cos\theta). \quad\text{..............(ii)}$$

Whence, multiplying (i) by $\dot{\psi}$, we get, at the two limiting positions when $\dot{\theta}=0$,

$$C\omega\dot{\psi}(\cos\theta_0-\cos\theta)=2Mga(\cos\theta_0-\cos\theta),$$

that is, either $\cos\theta=\cos\theta_0$, and hence from (i),

$$\dot{\psi}=0, \quad\text{.......................................}(\alpha)$$

$$\text{or}\quad C\omega\dot{\psi}=2Mga. \quad\text{..................................}(\beta)$$

Relation (a) gives the upper limit of oscillation, *i.e.* the original position at which the top was spun (see Art. 98).

Relation (β) gives the value of the azimuthal velocity $\dot{\psi}$ at the lower limit of oscillation.

101. To determine the least value of ω which will enable a top to spin, if set down at an inclination θ_0 to the vertical. We have seen that the top will first fall to a position θ_1 and then rise to its original position; but there must be some angle θ_2 at which the top fails to spin, either through the peg slipping, namely the constraint at O giving way, or through the side of the top touching the ground. Hence the limiting value θ_1 must be reached before this value θ_2, and the necessary condition is $\theta_2 > \theta_1$, and therefore $(\dot{\psi})_2 > (\dot{\psi})_1$; for writing

$$\dot{\psi}=\frac{C\omega(1-\cos\theta-k)}{A(1-\cos^2\theta)}=\frac{C\omega}{A}\left\{\frac{1}{1+\cos\theta}-\frac{k}{1-\cos^2\theta}\right\}$$

we see that since k is a positive constant $\dot{\psi}$ increases with θ.

But $$(\dot{\psi})_2=\frac{C\omega(\cos\theta_0-\cos\theta_2)}{A\sin^2\theta_2} \quad \text{from (iii) Art. 98,}$$

and $$(\dot{\psi})_1=\frac{2Mga}{C\omega} \quad \text{from } (\beta) \text{ Art. 100.}$$

$$\therefore \frac{C\omega(\cos\theta_0-\cos\theta_2)}{A\sin^2\theta_2} > \frac{2Mga}{C\omega}$$

$$\text{or} \quad \omega^2 > \frac{2MgaA\sin^2\theta_2}{C^2(\cos\theta_0-\cos\theta_2)}.$$

Example. Suppose the radii of gyration of the top are respectively $\frac{5}{8}$ in. and 2 ins. about the axle and perpendicular axis through the point of support. Let the distance a between the centre of gravity and the toe be 2 ins.; let $\theta_0=30°$ and $\theta_2=60°$.

If ω is the least angular velocity required

$$\omega^2 > \frac{2MgaA\sin^2\theta_2}{C^2(\cos\theta_0-\cos\theta_2)}$$

$$> 2.386.2.4.\frac{64^2}{25^2}\left(\frac{\sin^2 60°}{\cos 30°-\cos 60°}\right),$$

since $g=386$ inch/sec^2;

$$\therefore \omega > \frac{4.64}{25}\cdot\sqrt{386}\times\sqrt{2{\cdot}049}$$

$$> 289 \text{ rad./sec. approx.}$$

· number of revolutions per sec. must be

$$> 47.$$

Thus if the top is thrown from a string coiled round it in circles of average radius 1 in., the hand must be withdrawn from the top at a relative velocity of about 16 mi./hr.

102. To determine the least velocity at which a top will spin in a vertical position. In this case we have from equation (iii) Art. 98,

$$\dot\psi = \frac{C\omega(1-\cos\theta)}{A\sin^2\theta}$$

$$= \frac{C\omega}{A(1+\cos\theta)}.$$

If now we suppose θ so small that θ^2 may be neglected, we see that

$$\dot\psi - \frac{C\omega}{2A}, \text{ a constant.}$$

Again, from equation (ii) Art. 98, the value of $\dot\psi$, when θ is stationary, is given by

$$\dot\psi^2 = \frac{2Mga(1-\cos\theta)}{A\sin^2\theta}$$

$$= \frac{2Mga}{A(1+\cos\theta)}$$

$$= \frac{Mga}{A},$$

neglecting squares of small quantities.

Hence, in order that the top may return to the vertical position after a small displacement, it is necessary that

$$\frac{C^2\omega^2}{4A^2} > \frac{Mga}{A},$$

$$\text{or} \quad C^2\omega^2 > 4MgaA.$$

When ω^2 is less than this value, the disturbed motion does not, as the disturbance is indefinitely diminished, tend to return to the vertical motion; on the other hand, it does not necessarily depart very far from it. Both limits of oscillation in agreement with ω^2 might be *near* the vertical.

The above conditions show that it is very difficult to spin a top for which C is small in comparison with A; for instance it is almost impossible to spin a pencil on its point.

Example. Taking the top of the previous article we must have

$$\omega^2 > \frac{4MgaA}{C^2}$$

$$> \frac{4 \cdot 386 \cdot 2 \cdot 4 \cdot 64^2}{25^2},$$

$$\omega > \frac{8 \cdot 64}{25}\sqrt{193}$$

$$> 284{\cdot}5 \text{ rad./sec.}$$

· the number of revolutions a second must be

$$> 45 \text{ approx.},$$

which is practically the same number as in the preceding case.

103. Change of sign of $\dot{\psi}$. Let the axle OH (Fig. 55) be taken to represent the angular momentum $C\omega$ about itself, and HL the component due at any instant to the resolute of $\dot{\psi}$. Then OL represents the total angular momentum about axes in the azimuthal plane; and if LM is drawn perpendicular to OZ it follows that OM represents the constant angular momentum about the vertical, since $\dot{\theta}$ has no resolute about OZ. *Hence L always lies on the fixed horizontal plane through M.* This is sometimes called the *invariable plane.* It is clear that if OH is less than OM, HL can never vanish. Consequently $\dot{\psi}$ does not vanish.

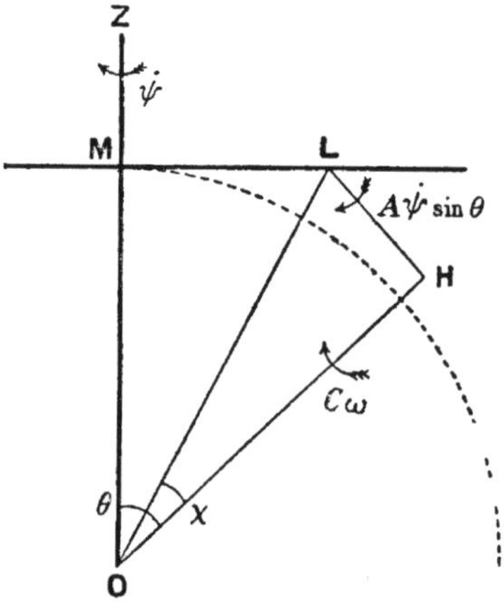

Fig. 55.

If OH is equal to OM, $\dot{\psi}$ vanishes when H reaches M, *i.e.* when the top becomes vertical. If OH is greater than OM, $\dot{\psi}$ vanishes when H reaches the invariable plane, and changes sign as H rises higher, since HL is then drawn in the opposite direction.

104. Gyroscopic resistance of the momentum axis. The existence of the limiting values of θ (Art. 99) can be seen from the following considerations.

We have seen (Art. 85), that if M is the angular momentum at any instant about OL, of a body to which is being applied a torque about a perpendicular axis, then the resistance the momentum-axis offers to being turned by the torque is $M\Omega$, where Ω is the precessional velocity already existing about the third axis perpendicular to both.

Now (Fig. 55), as the momentum-axis OL is tilted, M increases and also (Art. 85) the precession Ω; consequently the resisting torque increases. Suppose the top is spun initially with its axle at an angle θ_0 to the vertical: the axle of the top is initially the momentum-axis. As the gravity torque tilts this, the resisting torque is increased. At first it is less than the gravity torque; then a position is reached at which it is equal to the gravity torque, and at which steady motion would be set up if $\dot{\theta}$ were zero; and finally, as it increases and exceeds the gravity torque, a position, θ_1, is reached where $\dot{\theta}$ has been reduced to zero. The axle then starts to retrace its path in the azimuthal plane. In the same way, as it retraces its path, a position is reached where the gravity torque is equal to the resisting torque, which is tending to restore the more vertical position, and steady motion would ensue but for $\dot{\theta}$ (now a negative quantity); and finally the position $\theta=\theta_0$ is reached where $\dot{\theta}$ is again reduced to zero.

Since the angle for steady motion lies between θ_0 and θ_1, and the tendency is always to return towards it, the position of steady motion must be stable.

If there were a slight frictional couple at the toe to diminish ω, the limiting positions both of rise and fall would gradually descend until some position θ_2 for the lower limit were reached, at which the top either skidded at the point of support or touched the table with its side (see Art. 101). In practice the frictional forces which act on the top generally damp the oscillations so that they are hardly perceptible, and the top appears to descend steadily. See Chapters IV. and IX.

105. Value of the resisting torque at any instant. If χ be the angle which OL (Fig. 55) makes at any given instant with the axle of the top,

$$\tan\chi=\frac{A\dot{\psi}\sin\theta}{C\omega}$$
$$=\frac{\cos\theta_0-\cos\theta}{\sin\theta},$$

from equation (i), Art. 98.

The resisting torque

$$M\Omega=C\omega\sec\chi\dot{\psi}\sin(\theta-\chi)$$
$$=C\omega\dot{\psi}(\sin\theta-\cos\theta\tan\chi)$$
$$=C\omega\dot{\psi}(\sin\theta-\cot\theta\,\overline{\cos\theta_0-\cos\theta})$$
$$=\frac{C\omega\dot{\psi}}{\sin\theta}(1-\cos\theta_0\cos\theta),$$

or in terms of θ only

$$=\frac{C^2\omega^2(\cos\theta_0-\cos\theta)(1-\cos\theta_0\cos\theta)}{A\sin^3\theta}$$

from equation (iii), Art. 98.

From the value of $\tan\chi$ we obtain

$$\cos(\theta-\chi)=\cos\theta_0\cos\chi.$$

Now since θ and $\dot{\psi}\sin\theta$ increase together (Fig. 55), it follows that θ and χ increase together; consequently $(\theta-\chi)$ increases with θ. Hence both $C\omega\sec\chi$ and $\dot{\psi}\sin(\theta-\chi)$ have their maxi mum values at the position θ_1, and their minimum values at θ_0, showing that the resisting torque is a maximum at the lower limit of oscillation and a minimum at the upper limit.

It should be noticed that if C is small compared with A the resisting torque is very small. Compare Art. 102.

106. Equation of motion in the azimuthal plane. It follows that the oscillations in the azimuthal plane are determined by the equation

$$A\ddot{\theta}=Mga\sin\theta-\frac{C^2\omega^2(\cos\theta_0-\cos\theta)(1-\cos\theta_0\cos\theta)}{A\sin^3\theta}.$$

Multiplying both sides by $2\dot{\theta}$ and integrating we get the first equation of Art. 99,

while $$\dot{\psi}=\frac{C\omega(\cos\theta_0-\cos\theta)}{A\sin^2\theta}.$$

Thus the motion of the axle in the azimuthal plane being known, and that of the azimuthal plane, the complete motion of the top is known in terms of θ.

If θ_0 is zero, and θ small, the above equation reduces to

$$A\ddot{\theta}=Mga\sin\theta-\frac{C^2\omega^2(1-\cos\theta)^2}{A\sin^3\theta},$$

$$A\ddot{\theta}=Mga\,\theta-\frac{C^2\omega^2}{4A}\theta,$$

neglecting θ^2 and higher powers,

or $$\ddot{\theta}+\left(\frac{C^2\omega^2-4MgaA}{4A^2}\right)\theta=0,$$

while

$$\dot{\psi}=\frac{C\omega}{2A}=\omega_1 \text{ say. (Art. 102.)}$$

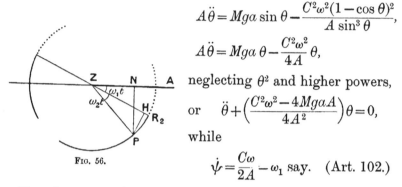

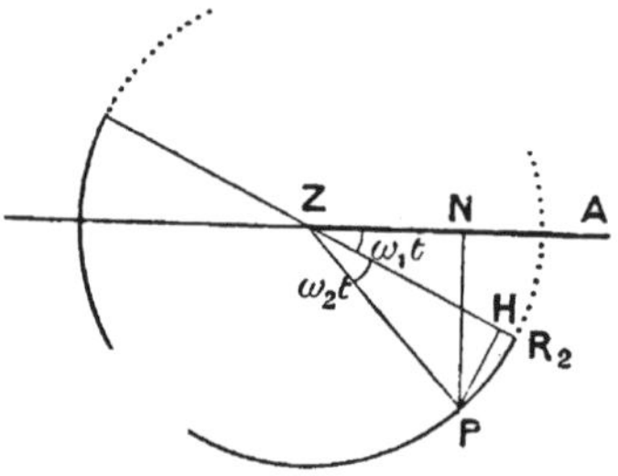

FIG. 56.

The above equations show that the axle oscillates about the vertical with a simple harmonic motion of period $\frac{2\pi}{\omega_2}$ where

$$\omega_2=\sqrt{\frac{C^2\omega^2-4MgaA}{4A^2}},$$

and the condition for a real oscillation is the condition for stability obtained in Art. 102.

Since H the head of the top (Fig. 56) describes a simple harmonic motion in ZR_2 the azimuthal plane, while ZR_2 rotates uniformly with angular velocity ω_1, the velocity of H at any moment is known, and its path in space completely determined. Considering the component velocities of H in the directions HZ, HP, we see that, when viewed from a point in the axis of Z, H will appear to describe a series of loops as in Fig. 63 in periodic time $\frac{2\pi}{\omega_2}$.

It will be noticed that the point N oscillates in the fixed plane ZA in periodic time $\frac{2\pi}{\omega_1+\omega_2}$. (See Art. 141.)

107. Motion of the top when G has an initial velocity. The principles established in Art. 104 show that, whatever be the initial velocity of G, the top still oscillates between two limiting positions. These limits can be found analytically as in Art. 99, by eliminating $\dot{\psi}$ between the two equations for conservation of angular momentum and conservation of energy.

The condition that $\dot{\theta}$ should be zero gives a cubic in $\cos\theta$, having one real root between $\cos\theta=-1$ and $\cos\theta=\cos\theta_0$, and another real root between $\cos\theta=\cos\theta_0$ and $\cos\theta=1$. The third root is inadmissible, being greater than unity.

If D is the initial angular momentum about the vertical, which remains constant throughout the motion, we have in this case

$$A\dot{\psi}\sin^2\theta+C\omega\cos\theta=D,$$

or, if we write $D=C\omega b\cos\theta_0$ where b and θ_0 are constants,

$$\dot{\psi}=\frac{C\omega(b\cos\theta_0-\cos\theta)}{A\sin^2\theta}.$$

Also $$\tan\chi=\frac{A\dot{\psi}\sin\theta}{C\omega}$$

$$=\frac{b\cos\theta_0-\cos\theta}{\sin\theta}.$$

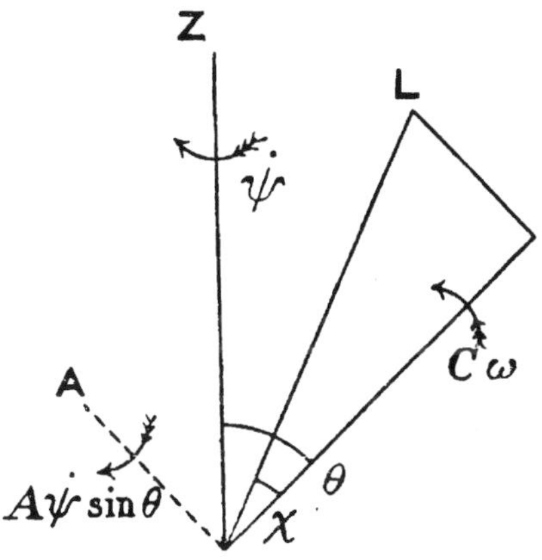

FIG. 57.

Hence the resisting torque becomes

$$\frac{C^2\omega^2(b\cos\theta_0-\cos\theta)(1-b\cos\theta_0\cos\theta)}{A\sin^3\theta},$$

and the oscillations in the azimuthal plane are given by

$$A\ddot{\theta}=Mga\sin\theta-\frac{C^2\omega^2(b\cos\theta_0-\cos\theta)(1-b\cos\theta_0\cos\theta)}{A\sin^3\theta},$$

the integral of which equation gives the equation of energy.

108. Path in space of H, the head of the top. We can now discuss the general appearance of the path of H in space when viewed from the point Z vertically above the origin. The paths will vary according to the initial conditions of motion; but in all cases, where ω is sufficiently large to prevent the top from falling altogether, the head will oscillate between the two positions R_1, R_2, in the azimuthal plane, while the azimuthal plane itself rotates about the fixed vertical. The actual velocity of H in space is that resulting from these two motions.

Since $\dot{\psi}$ is a function of θ only, it is the same for all angular positions of H equidistant from Z.

It is also clear that H passes and returns through any one position in the azimuthal plane with the same velocity.

We will first consider the case in which G has no initial velocity.

Here the starting point S coincides with R_1, the upper limit of H. In Fig. 58 θ_0 is an acute angle, in Fig. 59 it is obtuse. Since G (and therefore H) has no initial velocity, H must in each case start in the invariable plane (Art. 103). In Fig. 60 θ_0, the upper limit, is zero.

Other cases can be represented in a similar manner. In Fig. 61, where θ_0 is acute, an initial velocity v has been given at S to the head of the top *in the azimuthal plane.* At this

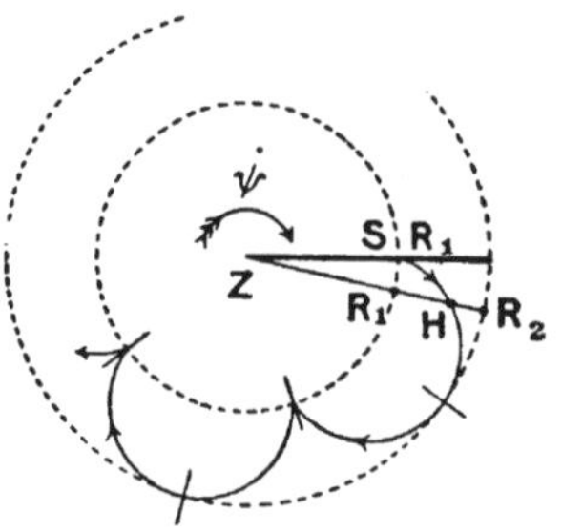

FIG. 58.

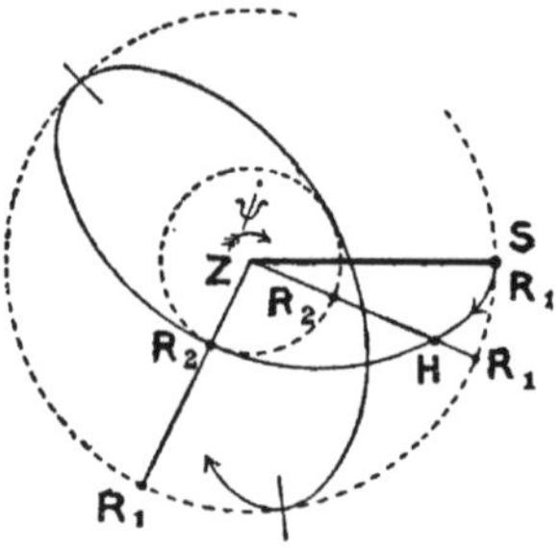

FIG. 59.

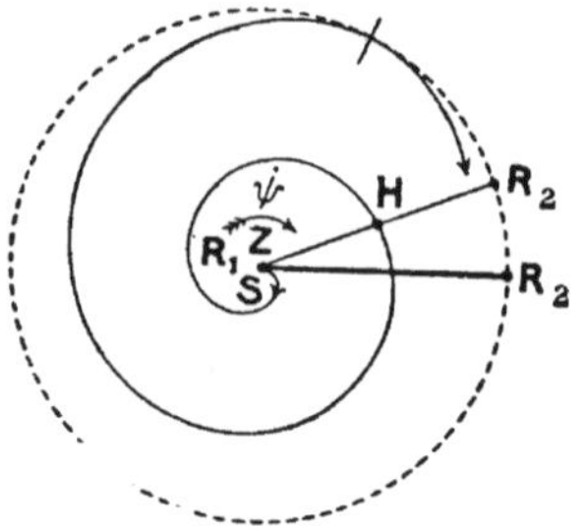

FIG. 60.

position $\dot{\psi}$ is zero, and therefore it will change sign as H returns with velocity v through the same position in the azimuthal plane. The path consequently has a loop as shown.

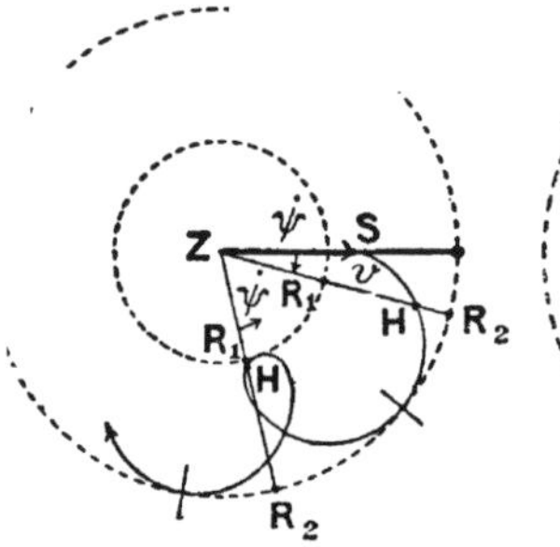

FIG. 61.

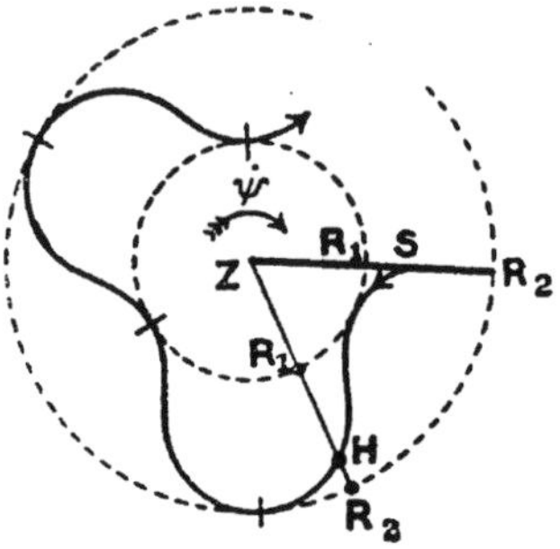

FIG. 62.

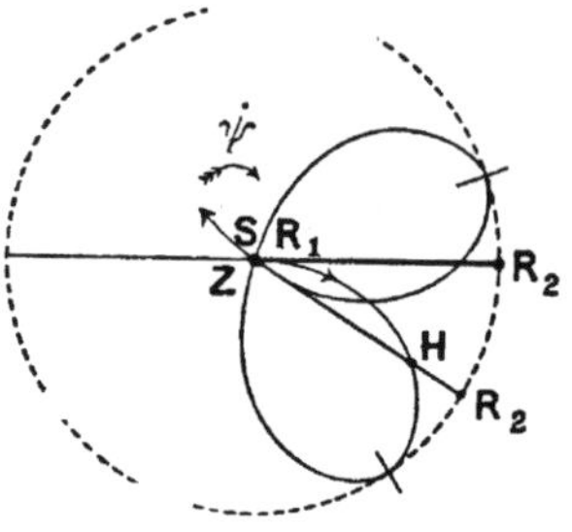

FIG. 63.

In Fig. 62 θ_0 is acute, and an initial velocity has been given *in the direction of precession,* causing the top to rise, though not necessarily, to a vertical position.

In Fig. 63 θ_0 is zero, but again an initial velocity v has been given to H in any direction. H therefore returns through its starting point with the same velocity, thus describing a series of loops.

109. Motion of any solid of revolution spinning on a smooth horizontal plane. Considering the case where G has no initial motion, we see that since the plane is smooth, the spin ω of the body about its axle remains constant, and G only moves in a vertical line.

Let θ be the inclination to the vertical at any time of the axle of the body, $GN=z$, $GL=x$, $PL=y$ (Fig. 64).

We can consider the whole motion of the body as comprised independently of the translational motion of G and the rotational motion about axes through G.

As before the angular momentum

$$\text{about } GC = C\omega,$$

$$\text{,, } \quad GA = A\dot{\psi}\sin\theta.$$

If the body is put down initially spinning with its axle at an angle θ_0 to the vertical, and G has no initial velocity, from conservation of angular momentum about the vertical we get

$$A\dot{\psi}\sin^2\theta + C\omega\cos\theta = C\omega\cos\theta_0,$$

$$\text{or} \quad \dot{\psi} = \frac{C\omega(\cos\theta_0 - \cos\theta)}{A\sin^2\theta} \text{ as before. } \qquad \text{(i)}$$

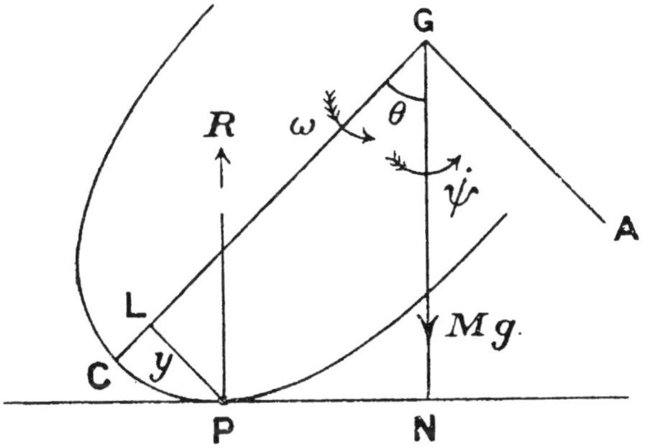

FIG. 64.

In considering the conservation of energy we must now include the energy of translation.

Let z_0 be the height of the centre of gravity above the plane originally: we have then

$$A\dot{\theta}^2 + A\dot{\psi}^2\sin^2\theta + M\dot{z}^2 = 2Mg(z_0 - z). \qquad \text{(ii)}$$

These two equations determine the whole motion, when combined with the geometrical relation

$$z = x\cos\theta + y\sin\theta,$$

x and y being functions of θ, depending on the nature of the meridian curve.

By reasoning exactly similar to that of Art. 104, it can be shown that GC will oscillate between two positions θ_0, θ_1, while G consequently oscillates between two positions G_0, G_1.

At the limits of oscillation

$$\dot{z} = 0, \quad \dot{\theta} = 0, \text{ and therefore from (ii),}$$

$$A\dot{\psi}^2\sin^2\theta = 2Mg(z_0 - z),$$

whence from (i),

$$C\omega\dot{\psi}(\cos\theta_0 - \cos\theta) = 2Mg(z_0 - z).$$

110. By the method of gyroscopic resistance. The gyroscopic resistance of the momentum-axis will in this case be the same as in Art. 105. Writing this expression as K, the oscillations in the azimuthal plane are given by

$$A\ddot{\theta}=R\,.\,p-K,$$

where $PN=p$, and R is given by the equation

$$M\ddot{z}=R-Mg.$$

As before we have

$$\dot{\psi}=\frac{C\omega(\cos\theta_0-\cos\theta)}{A\sin^2\theta}.$$

CHAPTER VIII.

MOVING AXES.

111. We will now discuss more fully the motion of a body referred to moving axes, that is, to a "frame of reference" (Art. 81) which is moving in *any* given manner.

The motion of the body relative to the frame when combined with the motion of the frame itself will entirely determine the motion of the body.

112. We will first consider the motion of a point in a plane referred to *two* moving axes.

If P be a point x, y referred to two *fixed* rectangular axes OX, OY, then the entire motion of P is determined if we know x, y, and consequently

$$\frac{dx}{dt}, \frac{dy}{dt} \text{ and } \frac{d^2x}{dt^2}, \frac{d^2y}{dt^2},$$

or in the fluxional notation,

$$\dot{x}, \dot{y} \text{ and } \ddot{x}, \ddot{y}.$$

Now suppose that, O remaining fixed, the axes turn about O but still remain rectangular (Fig. 65). The motion of the point P is now entirely determined when we know x, y (and therefore $\dot{x}$, $\dot{y}$ and $\ddot{x}$, $\ddot{y}$) *relative* to the moving frame of reference, or moving axes, and the motion of the axes themselves. This latter motion is of course entirely rotational since O remains fixed; and although OX, OY are revolving it must be remembered that *at any instant they coincide with two definite, fixed, directions in space.* We shall proceed to determine expressions for the *total* velocities and accelerations of the point P in these two definite directions fixed in space.

113. Linear velocities in two dimensions. *Method I.* Let OX and OY represent the positions of the axes at the time t, the instant in question, and let OX' OY' represent their positions at the time $t+\delta t$.

Let $x(=ON)$ and $y(=PN)$ mark the position relative to OX, OY of the moving point P at the time t, and $x+\delta x(=ON')$,

$y+\delta y(=P'N')$ the position relative to the axes at the time $t+\delta t$ when they have arrived at the positions OX', OY'.

Then if u and v are the component velocities at the time t in the fixed directions in space OX, OY we have

$$u=\underset{\delta t=0}{\text{Lt}}\frac{NL}{\delta t}$$

$$=\text{Lt}\frac{OL-ON}{\delta t}.$$

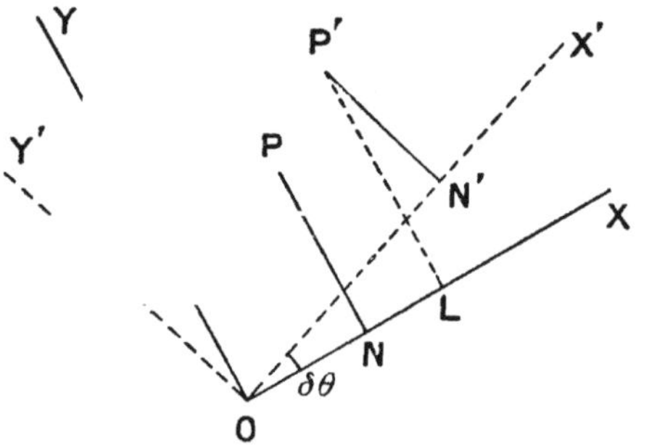

FIG. 65.

But $OL=$projection on OX of the broken line $ON'P'$

$$-(x+\delta x)\cos\delta\theta-(y+\delta y)\sin\delta\theta$$

$$-x+\delta x-y\delta\theta,$$

neglecting infinitesimals of the second order.

Also $$ON=x.$$

Hence, $$u=\frac{\delta x-y\delta\theta}{\delta t}$$

$$=\frac{dx}{dt}-y\frac{d\theta}{dt}\text{ in the limit,}$$

or $\dot{x}-y\theta$.

Similarly, $$v=\underset{\delta t=0}{\text{Lt}}\frac{\text{projection on }OY\text{ of }ON'P'-PN}{\delta t}$$

$$=\text{Lt}\frac{(y+\delta y)\cos\delta\theta+(x+\delta x)\sin\delta\theta-y}{\delta t}$$

$$=\frac{dy}{dt}+x\frac{d\theta}{dt},$$

or $\dot{y}+x\dot{\theta}$.

[30.] Test the dimensions of the above expressions.

114. *Method II.* These expressions may also be obtained in the following way:

$u=$velocity of P in direction OX

$=$velocity of P relative to N in this direction$+$that of N

$=-y\dot{\theta}+\dot{x}.$

And similarly,

v = velocity of P in direction OY

= velocity of P relative to N in this direction + that of N

$= \dot{y} + x\dot{\theta}$.

It will thus be seen that the total velocities of P in the directions OX, OY are *not merely* $\frac{dx}{dt}$ and $\frac{dy}{dt}$, but depend also on $\dot{\theta}$.

Cor. If $y=0$, x and θ become the polar coordinates of the point P. Hence, with the more usual notation, we see that

the radial velocity is $\dot{r}$,

and the transversal velocity is $r\dot{\theta}$.

115. Accelerations. We have above found u and v, the rates of change, in definite directions, of x and y, subject to certain conditions. To find the total accelerations in the same directions we must find the rates of change of u and v in these directions, subject to the same conditions.

Method I. The component of velocity in the direction OX after time δt is $(u+\delta u)\cos\delta\theta-(v+\delta v)\sin\delta\theta$.

Hence the acceleration in direction OX

$$=\underset{\delta t=0}{\mathrm{Lt}}\frac{(u+\delta u)\cos\delta\theta-(v+\delta v)\sin\delta\theta-u}{\delta t}$$

$$=\dot{u}-v\dot{\theta}.$$

Similarly, acceleration in direction $OY=\dot{v}+u\dot{\theta}$.

116. *Method II.* These expressions may also be obtained as follows: The velocities of the moving point P at any instant in the directions OX, OY, considered fixed in space, can be *represented* by $u=ON$ and $v=QN$,

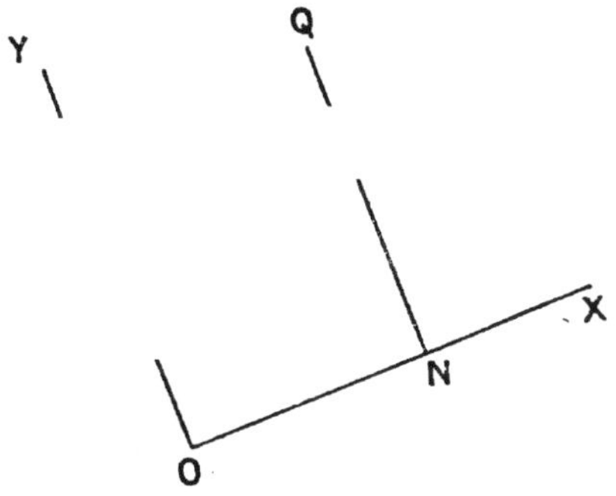

FIG. 66.

but it must be remembered that Q does not represent the *position* of the moving point.

The required accelerations will be the rates of change of u and v as the axes revolve, and as in Method II. of the previous article we found the rates of change of x and y to be

$$\dot{x}-y\dot{\theta} \text{ and } \dot{y}+x\dot{\theta},$$

so now the rates of change of u and v will be

$$\dot{u}-v\dot{\theta} \text{ and } \dot{v}+u\dot{\theta}.$$

Cor. It follows as in Art. 114 that, when polar coordinates are employed,

the radial acceleration is $\ddot{r}-r\dot{\theta}^2$,

and the transversal acceleration is $r\ddot{\theta}+2\dot{r}\dot{\theta}$ or $\frac{1}{r}\cdot\frac{d}{dt}(r^2\dot{\theta})$.

These results can also be obtained from first principles by a method similar to that employed in Art. 113.

117. It should be particularly noticed that in the previous articles the velocities and accelerations found are *total* velocities and accelerations—not relative to the moving frame of reference. Similar expressions can be obtained for a point referred to axes moving in three dimensional space, and for this purpose Method II. will be found most convenient.

118. Let x, y, z be the coordinates of the point P referred to three rectangular axes OX, OY, OZ (Fig. 67) which form a "frame of reference," while this frame turns with component angular velocities θ_1, θ_2, θ_3 about the lines OX, OY, OZ—or

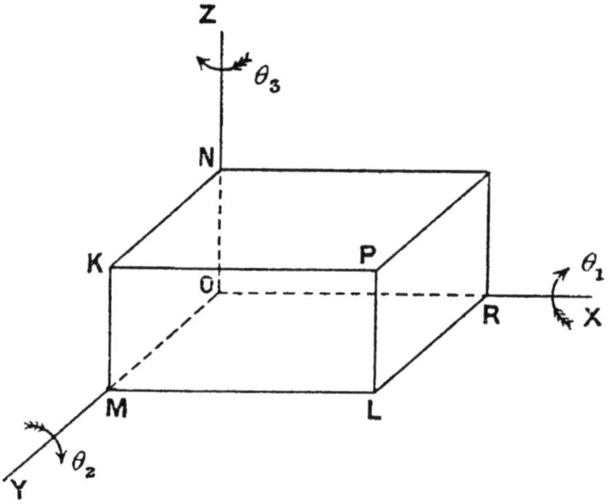

FIG 67.

rather about the lines fixed in space with which OX, OY, OZ happen at the instant under consideration to be coinciding.

We will consider that the point O remains fixed since we know that if we are discussing the motion of a rigid body and take our origin at the centre of gravity G of the body, then, as far as rotation about G is concerned, G is equivalent to a fixed point.

We will also adopt the following convention for the algebraic signs of rotation about OX, OY, OZ:

Rotation about	OX	is positive in	the	direction	Y to Z,
"	OY	"	"	"	Z to X,
"	OZ	"	"	"	X to Y,

namely, the positive direction is that of a left-handed screw as seen from O when looking along the axis; or, is determined by taking the cyclic change of the letters X, Y, Z.

119. Linear velocities in three dimensions. If u, v, w are the (total) component velocities in the directions of the axes, we have (Fig. 67)

u = component velocity of P relative to K + that of K relative to N + that of N relative to O,

and $$u = \dot{x} - y\theta_3 + z\theta_2,$$

and, similarly, by considering relative volocities,

$$v = \dot{y} - z\theta_1 + x\theta_3,$$
$$w = \dot{z} - x\theta_2 + y\theta_1.$$

It should be noticed that the dimensions of the expressions are correct.

120. Accelerations. Again, if we take OR, RL, LP (Fig. 67) to represent these velocities u, v, w, the rates of change of u, v, w, *i.e.* the total accelerations, along OX, OY, OZ are given by

$$\dot{u} - v\theta_3 + w\theta_2,$$
$$\dot{v} - w\theta_1 + u\theta_3,$$
$$\dot{w} - u\theta_2 + v\theta_1.$$

121. Angular velocities. Since angular velocity and angular momentum are vector quantities, the foregoing results are also applicable to them.

Let ω_1, ω_2, ω_3 be the (total) angular velocities of a rigid body about the axes OX, OY, OZ; that is, about the lines fixed in space with which OX, OY, OZ happen to be coinciding at the instant under consideration: it follows that the rates of change of the angular velocities about the moving axes are

$$\dot{\omega}_1 - \omega_2\theta_3 + \omega_3\theta_2,$$
$$\dot{\omega}_2 - \omega_3\theta_1 + \omega_1\theta_3,$$
$$\dot{\omega}_3 - \omega_1\theta_2 + \omega_2\theta_1.$$

122. Angular momentum. If O is a fixed point in a body referred to the three moving axes OX, OY, OZ, and h_1, h_2, h_3 are at any instant the components of the angular momentum of the body about these axes, then the rates of change of the angular momenta are given by

$$\dot{h}_1 - h_2\theta_3 + h_3\theta_2,$$
$$\dot{h}_2 - h_3\theta_1 + h_1\theta_3,$$
$$\dot{h}_3 - h_1\theta_2 + h_2\theta_1.$$

123. General equations of motion of a body having one point O fixed. Since the rate of change of angular momentum about an axis is equal to the moment of the external couple about that axis, it follows that if K_1, K_2, K_3 be the external couples acting about OX, OY, OZ, we get the following equations for determining the motion of a body·

$$\dot{h}_1 - h_2\theta_3 + h_3\theta_2 = K_1,$$
$$\dot{h}_2 - h_3\theta_1 + h_1\theta_3 = K_2,$$
$$\dot{h}_3 - h_1\theta_2 + h_2\theta_1 = K_3.$$

The equations of motion in the form thus obtained were first given by Mr. R. B. Hayward, F.R.S., of St. John's College, Cambridge. They are contained in a paper, published in 1856, in Part I. Vol. X. of the *Cambridge Philosophical Transactions*.

124. Euler's dynamical equations. In the case of the motion of a single rigid body about a fixed point, or about its centre of gravity, if we take for our moving axes three straight lines fixed in the body, and passing through the fixed point, or through the centre of gravity, we shall have

$$\theta_1 = \omega_1, \quad \theta_2 = \omega_2, \quad \theta_3 = \omega_3.$$

If, further, we take these three lines to be the principal axes at the fixed point,

$$h_1 = A\omega_1, \quad h_2 = B\omega_2, \quad h_3 = C\omega_3,$$

and the above equations reduce to

$$A\dot{\omega}_1 - (B - C)\omega_2\omega_3 = K_1,$$
$$B\dot{\omega}_2 - (C - A)\omega_3\omega_1 = K_2,$$
$$C\dot{\omega}_3 - (A - B)\omega_1\omega_2 = K_3,$$

which are Euler's equations.

125. General equations of motion of a body deduced from gyroscopic resistance. We have seen in Art. 37 that if angular momentum $I\omega$ about OX is precessing about OZ with velocity Ω under the action of a torque K about OY, then the gyroscopic resistance offered by the body to the torque K is measured by $I\omega\Omega$. Further, it will be remembered (Art. 38) that the direction of precession is determined by the angular momentum setting itself *towards* the torque-axis when both are drawn in the same sense.

In Fig. 68 let us suppose the system to be in motion under the action of external couples K_1, K_2, K_3 about OX, OY, OZ respectively.

It is clear that the angular momentum

h_1 is precessing about OY with velocity θ_2,

and „ „ OZ „ θ_3.

Similarly for the angular momenta h_2 and h_3.

It follows that when the external couple K_2 acts about OY the resistance offered is

due to h_1, measured by $h_1\theta_3$,

and „ h_3, „ $-h_3\theta_1$,

the negative sign being taken, since h_3 sets itself *away* from the torque-axis.

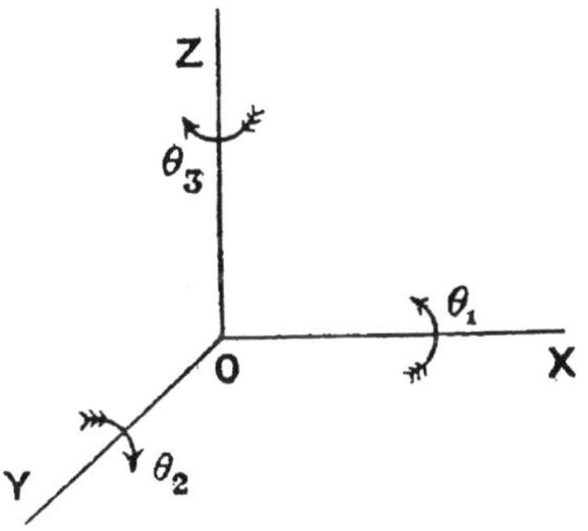

FIG. 68.

Hence, from moments about OY, we get

$$K_2 - h_1\theta_3 + h_3\theta_1 = \dot{h}_2,$$
$$\text{or} \quad K_2 = \dot{h}_2 - h_3\theta_1 + h_1\theta_3,$$

and two similar expressions as before.

126. It will be seen that our fundamental equation $K = I\omega\Omega$ (Art. 37) is a special case of the general equations, namely, when

$$\theta_1 = \theta_2 = 0, \quad \theta_3 = \Omega,$$
$$h_1 = I\omega, \text{ constant}, \quad h_2 = 0, \quad h_3 = C\Omega,$$
$$K_1 = K_3 = 0, \quad K_2 = K.$$

For in this case the first and third of the general equations vanish, and the second becomes

$$h_1\theta_3 = K_2,$$
$$\text{or} \quad I\omega\Omega = K.$$

127. Applications of moving axes.

Steady motion of a top on a fine fixed point. We will now employ the above expressions to determine the motion of a top spinning steadily as in Fig. 69, taking moving axes as marked in the figure, which is entirely in the azimuthal plane, axis (2) being perpendicular to (1) and (3) and towards the reader. We have

$$\theta_1 = -\Omega \sin\alpha \cdot \qquad \omega_1 = -\Omega\sin\alpha\,;$$
$$\theta_2 = \dot{\alpha} = 0\,; \qquad \omega_2 = 0\,;$$
$$\theta_3 = \Omega\cos\alpha \cdot \qquad \omega_3 = \omega.$$

Also $\qquad h_1 = A\omega_1\,; \quad h_2 = A\omega_2\,; \quad h_3 = C\omega_3.$

Referring to the equations of Art. 123, we see that the first vanishes, and the second one becomes

$$-C\omega_3\theta_1 + A\omega_1\theta_3 = K_2,$$

i.e. $C\omega\Omega \sin a - A\Omega \sin a\Omega \cos a = Mga \sin a$

or $C\omega\Omega - A\Omega^2 \cos a = Mga,$

the equation for steady motion already obtained. The third equation also vanishes.

We can in the same way obtain all the equations of steady motion established in Chapter VI.

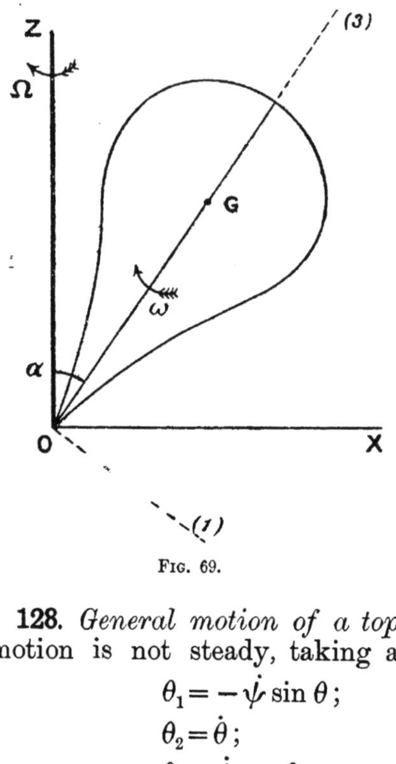

Fig. 69.

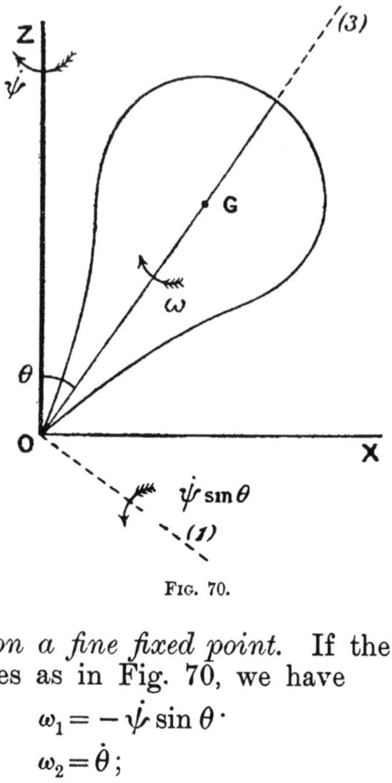

Fig. 70.

128. *General motion of a top on a fine fixed point.* If the motion is not steady, taking axes as in Fig. 70, we have

$$\theta_1 = -\dot{\psi} \sin \theta; \qquad \omega_1 = -\dot{\psi} \sin \theta \cdot$$
$$\theta_2 = \dot{\theta}; \qquad \omega_2 = \dot{\theta};$$
$$\theta_3 = \dot{\psi} \cos \theta; \qquad \omega_3 = \omega.$$

The equations of motion therefore are

$$A\dot{\omega}_1 - A\omega_2\dot{\psi} \cos \theta + C\omega_3\dot{\theta} = 0,$$
$$A\dot{\omega}_2 + C\omega_3\dot{\psi} \sin \theta + A\omega_1\dot{\psi} \cos \theta = Mga \sin \theta,$$
$$C\dot{\omega}_3 - A\omega_1\dot{\theta} - A\omega_2\dot{\psi} \sin \theta = 0.$$

Substituting the values of ω_1 and ω_2 in the third equation we see $C\dot{\omega}_3 = 0$, *i.e.* ω_3 is constant, a result which it will be seen depends on the moments of inertia of the top being the same about axes (1) and (2), *i.e.* on the symmetry of the top (see Art. 97, footnote).

The other two equations become

$$-A\ddot{\psi}\sin\theta - 2A\dot{\psi}\dot{\theta}\cos\theta + C\omega\dot{\theta} = 0, \quad \text{.................(i)}$$

$$A\ddot{\theta} + C\omega\dot{\psi}\sin\theta - A\dot{\psi}^2\sin\theta\cos\theta = Mga\sin\theta, \quad \text{......(ii)}$$

which completely determine the motion.

Multiplying the first of these by $\sin\theta$ and integrating we get

$$A\dot{\psi}\sin^2\theta + C\omega\cos\theta = D, \text{ a constant,}$$

which is the equation for conservation of angular momentum.

Again, multiplying (i) by $2\dot{\psi}\sin\theta$, and (ii) by $2\dot{\theta}$, subtracting (i) from (ii) and integrating

$$A\dot{\psi}^2\sin^2\theta + A\dot{\theta}^2 = E - 2Mga\cos\theta,$$

which is the equation of energy.

It will be noticed that of the three equations of motion that about axis (1) gives the condition for constancy of angular momentum about the vertical, that about axis (2) gives the oscillations in the azimuthal plane, and that about axis (3) the constancy of the spin of the top about its axle.

In a similar way the equations may be obtained for the general motion of a solid of revolution spinning on a smooth horizontal plane; but the shortest method is that of writing down the conditions for conservation of angular momentum and conservation of energy.

129. The following is typical of the more advanced problems to be found among the Miscellaneous Examples at the end of Chapter IX.

A rough horizontal plane is made to rotate about a fixed vertical axis with constant angular velocity, and the centre of a sphere lying at rest at the point where the axis meets the plane is set in motion with a given horizontal velocity. Show that the path of the centre in space is a circle, described with uniform velocity.

Let P be the point of contact at any instant of the sphere with the plane, O the point where the vertical axis OZ meets the plane. Taking OPX as axis (1), OZ as axis (3), and OY perpendicular to these as axis (2), we have $\theta_1 = 0$, $\theta_2 = 0$, and $\theta_3 = \theta$, where θ is the angle OP makes with some horizontal line fixed in space.

Let F_1, F_2 be the frictional forces at the point P in directions (1) and (2), f_1, f_2 the accelerations of the centre of gravity G in these directions, ω_1, ω_2, ω_3 the angular velocities of the sphere, m and a the mass and radius respectively.

Taking moments about axes through G parallel to (1) and (2), we have

$$\tfrac{2}{5}ma^2\dot{\omega}_1 - \tfrac{2}{5}ma^2\omega_2\dot{\theta} = F_2 \cdot a = maf_2, \quad \text{.................(i)}$$

and $$\tfrac{2}{5}ma^2\dot{\omega}_2 + \tfrac{2}{5}ma^2\omega_1\dot{\theta} = -F_1 a = -maf_1 \quad \text{...............(ii)}$$

Also if u and v are the velocities of G in the directions (1) and (2), by Art. 115

$$f_1 = \dot{u} - v\dot{\theta}; \quad f_2 = \dot{v} + u\dot{\theta}.$$

And, since there is no slipping, the point P of the sphere has the same velocity as the point P of the plane, whence

$$a\omega_2 - u = 0, \text{ i.e. } a\omega_2 = u,$$

and $$a\omega_1 + v = x\Omega, \text{ i.e. } a\omega_1 = x\Omega - v,$$

where Ω is the constant angular velocity of the plane.

Substituting in equations (i) and (ii) we obtain

$$\tfrac{7}{5}\dot{v} + \tfrac{7}{5}u\dot{\theta} = \tfrac{2}{5}\dot{x}\Omega,$$

and $$\tfrac{7}{5}\dot{u} - \tfrac{7}{5}v\dot{\theta} = -\tfrac{2}{5}x\Omega\dot{\theta},$$

showing that $$f_2 = \tfrac{2}{7}\dot{x}\Omega,$$

and $$f_1 = -\tfrac{2}{7}x\Omega\dot{\theta}.$$

Taking r, θ as the polar coordinates of the point P, these equations may be written

$$\frac{1}{r}\frac{d}{dt}(r^2\dot{\theta}) = \tfrac{2}{7}\Omega\dot{r}, \quad \text{.........................(iii)}$$

and $$\ddot{r} - r\dot{\theta}^2 = -\tfrac{2}{7}\Omega r\dot{\theta}. \quad \text{.........................(iv)}$$

Integrating (iii) we obtain

$$r^2\dot{\theta} = \frac{\Omega}{7}r^2 + \text{a constant}.$$

Since $\dot{\theta} = 0$ when $r = 0$ the constant is zero and

$$\dot{\theta} = \frac{\Omega}{7}.$$

Substituting in (iv) $$\ddot{r} + \frac{\Omega^2}{49}r = 0.$$

Since $\dfrac{d\theta}{dt} = \dfrac{\Omega}{7}$, the above may be written

$$\frac{d^2r}{d\theta^2} + r = 0,$$

whence the path of P in space is given by

$$r = c\cos(\theta + a),$$

where c and a are constants, showing that it is a circle.

Taking the sum of the squares of $\dot{r}$ and $r\dot{\theta}$ derived from this equation, we see that the velocity is constant and equal to that of projection.

CHAPTER IX.

STABILITY OF ROTATION. PERIODS OF OSCILLATION.

130. We shall now consider some cases where a spinning body receives a slight displacement, discussing the conditions under which it will revert to the original motion when the small disturbing force is withdrawn, and the periodic time in which it will oscillate about its original position.

131. Proposition. *If OA, OB, OC, are the three principal axes (through a fixed point O) of a body whose moments of inertia about these axes are A, B, C (either in ascending or descending order of magnitude), then, for small displacements, rotation either about OA or OC is stable, but about OB unstable.*

For, considering that the disturbing force is instantly withdrawn, we have, by Euler's equations, since no forces act on the body,

$$A\dot{\omega}_1-(B-C)\omega_2\omega_3=0,$$
$$B\dot{\omega}_2-(C-A)\omega_3\omega_1=0,$$
$$C\dot{\omega}_3-(A-B)\omega_1\omega_2=0.$$

If the body was spinning before displacement with velocity Ω about the axis of ω_1, we have

$$\omega_1=\Omega+\text{small quantity},$$
$$\omega_2=\text{small quantity},$$
$$\omega_3=\text{small quantity}.$$

Hence, to the first order, $A\dot{\omega}_1=0$, and thus ω_1 is constant $=\Omega$,

$$\text{whence}\quad B\dot{\omega}_2-(C-A)\Omega\omega_3=0,$$
$$\text{and}\quad C\dot{\omega}_3-(A-B)\Omega\omega_2=0;$$
$$BC\ddot{\omega}_2+(A-C)(A-B)\Omega^2\omega_2=0, \quad \text{..............(i)}$$

and the motion is simple harmonic.

The time between two successive equal values of ω_2 is given by

$$\frac{2\pi}{\Omega}\sqrt{\frac{BC}{(A-C)(A-B)}},$$

which is a real quantity, and the motion is stable. Similarly if the spin Ω is about the axis of ω_3; but if it is about the axis of ω_2, the period becomes

$$\frac{2\pi}{\Omega}\sqrt{\frac{CA}{(B-A)(B-C)}},$$

which is imaginary.

Hence the motion is stable for rotation about OA or OC but unstable about OB.

132. Diabolo which will not spin. The reader will remember that when he is beginning to spin a Diabolo spool by means of the string, unless the two portions of the string are *absolutely in the same plane and that at right angles to the axle of the spool*, then, besides the spin about the axle which he means to communicate, there is also a residuum "wobble" about a transverse axis; but as he continues his spinning this wobble begins to disappear and the spool settles down to a steady rotation. The reason for this is that the axle of the spool is in general an axis of minimum moment of inertia (though if it is one of maximum moment the same argument applies) and the instantaneous axis tends to revert to the position of the axle of the spool. If we construct a spool which is dynamically equivalent to a sphere we shall find that it is impossible to spin it; for, since the moments of inertia about all axes are equal, the spool has no tendency to revert to its axle as instantaneous axis of rotation. The result is that the axle wanders about indefinitely in space, though with considerable rapidity.

If this Diabolo were *already spinning* about its axle, it would possess a certain degree of stability; for, the spin about the axle being large compared to that generated by any disturbing force about another axis, the instantaneous axis would deviate only to a very slight degree from the axle of the spool.* (See Art. 63, footnote.)

A heavy conical sheet projecting equally on either side of the vertex whose semi-vertical angle is equal to $\tan^{-1}\sqrt{2}$ (Fig. 71), has the dynamical properties of a sphere; but such an ideal construction is impracticable. Since, however, the addition of matter inside the angles AOB', $A'OB$ would increase the moment of inertia about the axis of the cone, while the addition of matter inside the angles AOB, $A'OB'$ would increase the moment of inertia about a transverse axis, we can on such a skeleton cone form a material double hollow cone, in the form of a Diabolo spool which is dynamically equivalent to a sphere; and this Diabolo we shall find almost impossible to spin.

* It is for this reason that in the game of Cup and Ball, when it is required to jerk the ball so that the axial hole catches on the wooden peg, it is necessary to first give the ball a spin about the axis containing the hole and the point where the string is fastened to the ball; for then a lateral jerk of the string does not appreciably displace the axis of rotation.

At a meeting of the Physical Society of London, in November, 1907, Mr. C. V. Boys exhibited some interesting experiments with a wooden spool of this description. His spool was pierced with an axial hole, in which condition the moment of inertia was *greater* about the axle than about a transverse axis. The spool was found easy to spin, and displayed considerable stability.

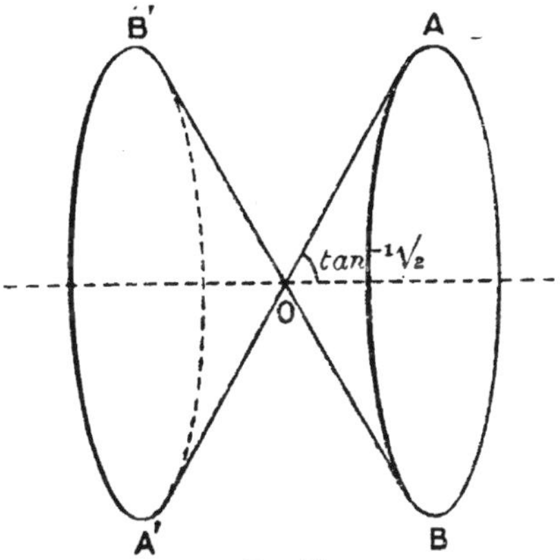

FIG. 71.

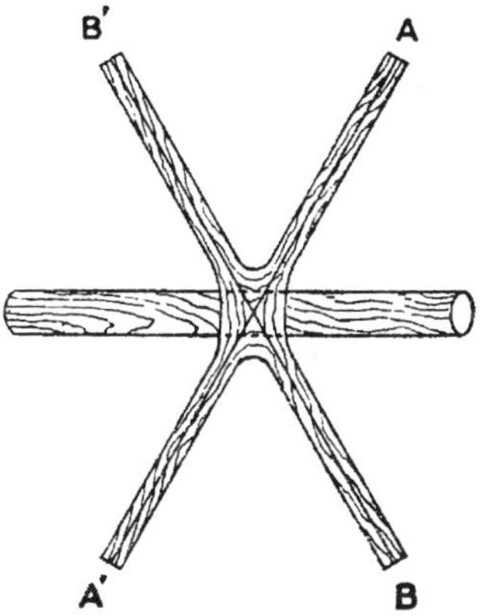

FIG. 72.

The lecturer then inserted through the hole (Fig. 72), a small stick of such a length as to make the two moments of inertia exactly equal, and by attempting to spin the spool illustrated very clearly its marked instability of rotation about any axis whatever, and the impossibility of spinning it. On the same stick being used but pushed slightly from its central position so as to project rather more at one end than the other, the spool again was spun without difficulty, and displayed distinct stability, but at once began to precess owing to the gravity couple called into play by the displacement of the centre of mass. A similar stability of rotation was observed when a longer stick was inserted, causing the moment of inertia to be *greater* about a transverse axis than about the axis of the cone.

It is interesting to notice that in equalising the moments of inertia the adjustment must be *very* exact before spinning becomes impossible. If a torsion wire is applied as a test it is found that when the difference in periods of oscillation is as much as 1 in 35, spinning is quite easy. The periods should not differ more than one per cent.

For several of the above details the author is indebted to the *Proceedings* of the above-mentioned meeting, published by the Society.

133. A problem of stability relating to Schlick's method of steadying vessels.

A fly-wheel is symmetrically mounted in a spherical case to which is fixed an axle perpendicular to the axis of the fly-wheel. The ends of this axle are mounted inside a hollow circular cylinder so that the direction of the axle intersects and is perpendicular to the axis of the cylinder. The cylinder lies on a rough table and the fly-wheel is set spinning with its axis vertical. Show that if a mass m be attached to the highest point of the cylinder, the system is stable, whatever be the mass of the cylinder, provided the square of the angular velocity of the fly-wheel exceeds $mgaA/C^2$, *where a is the outer radius of the cylinder, A the moment of inertia of the case and fly-wheel about the axle, and C that of the fly-wheel about its axis.*

Camb. Math. Tripos, 1908. (2nd Problem Paper.)

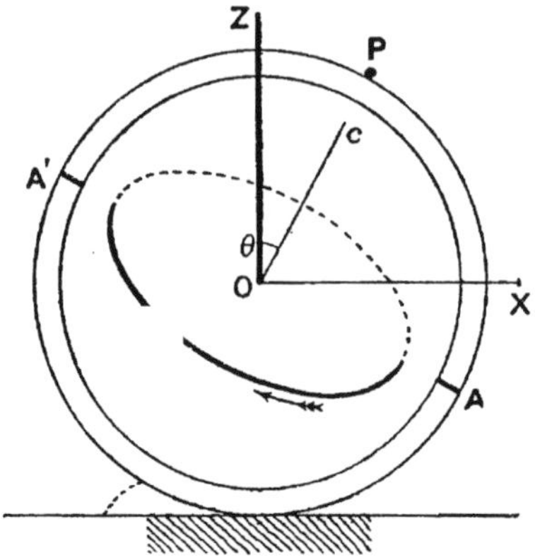

FIG. 73.

Let ZOX (Fig. 73) represent the vertical plane through the centre of the fly-wheel and the particle P, where the angle $ZOP=\theta$. After any time t let Oc be the projection on this plane of the axle OC of the fly-wheel, which is rotating in the direction marked, and has precessed through an angle ϕ towards the reader. If OB is perpendicular to OC and OA, then OB, originally coincident with the axis of the cylinder, now makes an angle ϕ with it.

Taking OA, OB, OC as moving axes, and considering the motion of the fly-wheel and case, we have

$$\omega_1=-\dot{\phi}, \quad \text{and} \quad \theta_1=-\dot{\phi};$$
$$\omega_2=\dot{\theta}\cos\phi, \quad \theta_2=\dot{\theta}\cos\phi;$$
$$\omega_3=\omega, \quad \theta_3=0.$$

Let K be the torque about the axis of the cylinder exerted on it by the fly-wheel and case, owing to the reactions at A and A'.

Taking moments about OA for the fly-wheel and case we have

$$\dot{h}_1-h_2\theta_3+h_3\theta_2=0,$$

or $$-A\ddot{\phi}+C\omega\dot{\theta}\cos\phi=0.$$

If the displacement is small, neglecting squares of small quantities,

$$\cos\phi=1,$$

and therefore $$A\dot{\phi}=C\omega\theta. \qquad \text{(i)}$$

Moments about OB give

$$\dot{h}_2-h_3\theta_1+h_1\theta_3=K\cos\phi,$$

$$\frac{d}{dt}(B\dot{\theta}\cos\phi)+C\omega\dot{\phi}=K\cos\phi,$$

or, neglecting squares of small quantities, and combining with (i)

$$B\ddot{\theta}+\frac{C^2\omega^2}{A}\theta-K=0. \qquad \text{(ii)}$$

Now, considering the motion of the cylinder, and taking moments about the generating line in contact with the table (moment of inertia I for cylinder and particle)

$$I\ddot{\theta}=mga\cos\phi\,.\,\sin\theta-K,$$

whence from (ii) we have, to the first order,

$$(B+I)\ddot{\theta}+\left(\frac{C^2\omega^2}{A}-mga\right)\theta=0,$$

so that the condition for stability is

$$\omega^2>\frac{mgaA}{C^2},$$

and the period of a small oscillation is

$$2\pi\sqrt{\frac{A(B+I)}{C^2\omega^2-mgaA}}.$$

134. Oscillations of a spinning top.

Case 1. *To find the period of a small oscillation of a spinning top about the position of steady motion where the axle is inclined at an angle α to the vertical.*

With the preceding notation we have, from conservation of angular momentum,

$$C\omega(\cos\alpha-\cos\theta)=A\dot{\psi}\sin^2\theta-A\Omega\sin^2\alpha. \qquad \text{(i)}$$

Writing down the equations of motion about the torque axis we have, including gyroscopic resistances,

$$A\ddot{\theta}=\sin\theta(Mga-C\omega\dot{\psi}+A\dot{\psi}^2\cos\theta), \qquad \text{(ii)}$$

while

$$Mga-C\omega\Omega+A\Omega^2\cos\alpha=0. \qquad \text{(iii)}$$

If θ becomes $\alpha+\epsilon$ and consequently $\dot{\psi}$ becomes $\Omega+\dot{\chi}$ where ϵ and $\dot{\chi}$ are small, we have

$$\sin\theta=\sin(\alpha+\epsilon)=\sin\alpha+\cos\alpha\,.\,\epsilon,$$

$$\cos\theta=\cos(\alpha+\epsilon)=\cos\alpha-\sin\alpha\,.\,\epsilon,$$

neglecting squares of small quantities.

Hence, from (i), $\quad C\omega(\cos\alpha-\cos\theta)=C\omega\sin\alpha\,.\,\epsilon,$

and

$$A\dot{\psi}\sin^2\theta-A\Omega\sin^2\alpha=A(\Omega+\dot{\chi})(\sin^2\alpha+2\sin\alpha\cos\alpha\,.\,\epsilon)-A\Omega\sin^2\alpha$$
$$-2A\Omega\sin\alpha\cos\alpha\,.\,\epsilon+A\dot{\chi}\sin^2\alpha;$$

$$\therefore\ A\dot{\chi}\sin\alpha=(C\omega-2A\Omega\cos\alpha)\epsilon$$
$$=\frac{Mga-A\Omega^2\cos\alpha}{\Omega}\,.\,\epsilon, \quad \text{from (iii)}.$$

Making the same substitution in (ii) we have

$$A\ddot{\epsilon}=[\sin\alpha+\cos\alpha\,.\,\epsilon][Mga-C\omega(\Omega+\dot{\chi})$$
$$+A(\Omega^2+2\Omega\dot{\chi})(\cos\alpha-\sin\alpha\,.\,\epsilon)]$$
$$=-\dot{\chi}\sin\alpha(C\omega-2A\Omega\cos\alpha)-A\Omega^2\sin^2\alpha\,.\,\epsilon,\text{ employing (iii)},$$
$$=-\epsilon\left[\frac{(Mga-A\Omega^2\cos\alpha)^2}{A\Omega^2}+A\Omega^2\sin^2\alpha\right],$$

i.e. $$\ddot{\epsilon}+\left[\frac{A^2\Omega^4-2MgaA\Omega^2\cos\alpha+M^2g^2a^2}{A^2\Omega^2}\right]\epsilon=0;$$

the period is $\dfrac{2\pi}{p}$,

where $$p=\frac{1}{A\Omega}\sqrt{A^2\Omega^4-2MgaA\Omega^2\cos\alpha+M^2g^2a^2}.$$

Since the quantity under the radical can be written

$$(A\Omega^2-Mga\cos\alpha)^2+(Mga\sin\alpha)^2,$$

it is always positive, and we see that the position of steady motion is stable, a result already established in Art. 104.

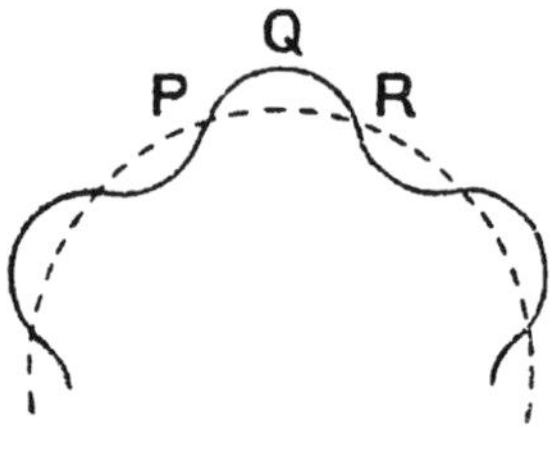

FIG. 74.

It should be noticed that the node P (Fig. 74) travels uniformly round the circle of steady motion with angular velocity Ω;

$\therefore$ the periodic time is $\dfrac{PR}{\Omega}=\dfrac{2\pi}{p}$;

$\therefore$ the length of $$PR=\frac{2\pi\Omega}{p}$$
$$=\frac{2\pi A\Omega^2}{\sqrt{A^2\Omega^4-2MgaA\Omega^2\cos\alpha+M^2g^2a^2}}.$$

These oscillations have already been mentioned in Chapter IV., where they were accounted for by the inertia of the top alternately resisting and hurrying the precession. Where the disturbing force is being continually increased, as for instance when a top is gradually sinking, these oscillations are not regular but instantaneous only about the instantaneous position of steady motion.

135. Case 2. The sleeping top.

Definition. A top is said to be asleep when the whole motion of its axle is in the neighbourhood of the vertical, *but not necessarily passing through it.*

If a small displacement be given to a sleeping top, two independent vibrations are set up, namely the oscillations of θ and $\dot{\psi}$ about their mean values; but since $\dot{\psi}$ is a function of θ only, it follows that the periods of these two oscillations are the same and the mean values occur simultaneously. The remaining Articles will be devoted to the discussion of this period and the values of θ, $\dot{\psi}$ and ψ at any time during the motion.

136. To find the value of θ at any instant. The equations of conservation of angular momentum and energy give (Art. 107),

$$A\dot{\psi}\sin^2\theta + C\omega\cos\theta = C\omega b\cos\theta_0, \quad \text{(i)}$$

and

$$A\dot{\psi}^2\sin^2\theta + A\dot{\theta}^2 + 2Mga\cos\theta = E. \quad \text{(ii)}$$

Putting $\frac{C\omega}{2A} = \lambda_1$, and $b = 1 + \frac{1+\beta}{2}.\theta_0^2$ where β is a constant, the former becomes $A\dot{\psi}\theta^2 = \frac{C\omega}{2}(\theta^2 + \beta\theta_0^2)$ neglecting cubes and higher powers of θ,

$$\textit{i.e.}\quad \dot{\psi} = \lambda_1\left(1 + \frac{\beta\theta_0^2}{\theta^2}\right), \quad \text{(iii)}$$

and the initial value of $\dot{\psi} = \lambda_1(1+\beta)$.

Hence we see that β is the ratio of the initial excess of $\dot{\psi}$ over λ_1 to λ_1.

It should be noticed that if θ_0 is zero, $\dot{\psi} = \lambda_1$, and is therefore constant, as was shown in Art. 102.

Combining (ii) with (iii) we obtain

$$\frac{\lambda_1^2(\theta^2 + \beta\theta_0^2)^2}{\theta^2} + \dot{\theta}^2 - \frac{Mga\theta^2}{A} = \text{constant},$$

reducing to

$$\dot{\theta}^2 + \lambda_2^2\theta^2 + \frac{\beta^2\theta_0^4\lambda_1^2}{\theta^2} = \mu, \text{ a constant},$$

$$\text{where}\quad \lambda_2^2 = \lambda_1^2 - \frac{Mga}{A},$$

and the condition for λ_2 being real is the condition for stability obtained in Art. 106.

If $\theta^2 = x$, this becomes

$$\frac{\dot{x}^2}{4} + \lambda_2^2x^2 + \beta^2\theta_0^4\lambda_1^2 - \mu x = 0, \quad \text{(iv)}$$

$$\text{whence}\quad 2\lambda_2 t + 2\delta = \int \frac{dx}{\left(\frac{\mu x}{\lambda_2^2} - \frac{\beta^2\theta_0^4\lambda_1^2}{\lambda_2^2} - x^2\right)^{\frac{1}{2}}},$$

$$\left.\begin{aligned} &\text{or}\quad \theta^2 = x = K' + k'\cos(2\lambda_2 t + 2\delta), \\ &\text{or}\quad \left(\frac{\theta}{\theta_0}\right)^2 = K + k\cos(2\lambda_2 t + 2\delta), \end{aligned}\right\} \quad \text{(v)}$$

where K, k, and δ are constants. (Williamson, *Integral Calculus*, p. 12.)

It follows that the period of oscillation of θ and $\dot{\psi}$ about their mean values is $\frac{\pi}{\lambda_2}$.

137. To determine the constants K, k, and δ in terms of θ_0 and $\dot{\theta}_0$.

From equation (v) we have

$$\dot{x} = -2\lambda_2 k\theta_0^2 \sin(2\lambda_2 t + 2\delta);$$

$$\therefore \frac{\dot{x}^2}{4} + \lambda_2^2(x - K\theta_0^2)^2 = \lambda_2^2 k^2 \theta_0^4.$$

Comparing this with equation (iv) with which it must be identical we get, by equating the constant terms,

$$(K^2 - k^2)\lambda_2^2 = \beta^2\lambda_1^2, \quad \text{.....................(vi)}$$

showing that K is always numerically $> k$.

Writing

$$\frac{\beta^2\lambda_1^2}{\lambda_2^2} = \gamma^2,$$

we have

$$K^2 - k^2 = \gamma^2, \quad \text{.......................(vii)}$$

and from (v)

$$K + k\cos 2\delta = 1, \quad \text{.......................(viii)}$$

while

$$k\sin 2\delta = -\frac{\dot{\theta}_0}{\theta_0\lambda_2} = \alpha \text{ say.} \quad \text{.....................(ix)}$$

These three equations determine K, k, and 2δ.

From (viii) and (ix) $k^2 = (1-K)^2 + \alpha^2$;

$\therefore$ using (vii) $K^2 - (1-K)^2 - \alpha^2 = \gamma^2$.

$$\therefore K = \frac{1+\alpha^2+\gamma^2}{2}.$$

Also

$$\tan 2\delta = \frac{2\alpha}{1-\alpha^2-\gamma^2}$$

and

$$k^2 = \left(\frac{1-\alpha^2-\gamma^2}{2}\right)^2 + \alpha^2.$$

From the last two equations we see that k may be either positive or negative, and also that δ can have two values. If k is taken as positive, we see from equation (ix) that δ must have that value which makes $\sin 2\delta$ of the same sign as α. Hence there is no ambiguity.

From equation (v) we see that the greatest and least values of $\left(\frac{\theta}{\theta_0}\right)^2$ are $K+k$ and $K-k$, and from (vi) the latter limit is always positive; hence θ never changes sign but varies between α_1 and α_2, where α_1, α_2 are $\theta_0\sqrt{K-k}$ and $\theta_0\sqrt{K+k}$ respectively.

138. Particular case when $K' = k'$. It should be remarked that this condition does not necessarily involve $K = k$, since θ_0 might be zero, which proves to be the case.

If $K'=k'$, and only in this case, equation (v) becomes

$$\begin{aligned}\theta^2 &= K'\{1+\cos(2\lambda_2 t+2\delta)\}\\ &= 2K'\cos^2(\lambda_2 t+\delta);\\ \therefore\ \theta &= \sqrt{2K'}\cos(\lambda_2 t+\delta),\end{aligned}$$

and the motion is simple harmonic.

In this case, since $K'^2-k'^2=\gamma^2\theta_0^{\,2}$,

we must have either $\theta_0=0$ or $\gamma=0$, which gives $\beta=0$.

The latter condition gives, from (iii), $\dot{\psi}=\lambda_1=\dfrac{C\omega}{2A}$.

Hence if $K'=k'$ the top is either initially vertical or is so displaced that the azimuthal velocity communicated to it is λ_1. The first of these two cases has already been considered in Art. 106; the second is clearly of the type discussed in Art. 134, a in this case being small. The path of the head H is a simple wave of period $\dfrac{2\pi}{\lambda_2}$. In Art. 140 it is shown that the path can in general be produced by the combined effect of two waves whose periods are $\dfrac{2\pi}{\lambda_1\pm\lambda_2}$, but it can be drawn without reference to these considerations by the method employed in the following Article.

139. The path in space of the head of the top as seen from above. We have seen (Art. 137) that θ never becomes zero, but oscillates between two limits a_1 and a_2.

If a graph be drawn such that the ordinates represent θ^2 and the abscissae the time, then equation (v) shows that the resulting curve is a simple wave of period $\dfrac{\pi}{\lambda_2}$.

If now we draw a second graph such that each ordinate is the square root of the corresponding one in the first curve, the new graph will give the values of θ for all time. This second curve is clearly flatter than the first, but repeats after the same period.

Now the value of ψ, if described with the mean value of $\dot{\psi}$, is proportional to the time; hence we may also regard this curve as representing θ in terms of the angular distance described with the mean value $\dot{\psi}$, and we are thus enabled to realise the general appearance of the path in space of the head H of the top.

The dotted curve sketched in Fig. 75 shows the path described on the assumption that $\dot{\psi}$ has its mean value throughout, while the continuous curve gives the actual path. The following considerations will show how they are connected.

Since $\dot{\psi}=\lambda_1+\frac{\beta\theta_0{}^2\lambda_1}{\theta}$, we see that when θ has its mean value, $\dot{\psi}$ also has its mean value; but as θ increases $\dot{\psi}$ diminishes, and *vice versâ*.

Let us suppose that the head H is at M when both θ and $\dot{\psi}$ have their mean values. This will again occur at M_1, M_2, etc., namely after successive periods of time $\frac{\pi}{2\lambda_2}$.

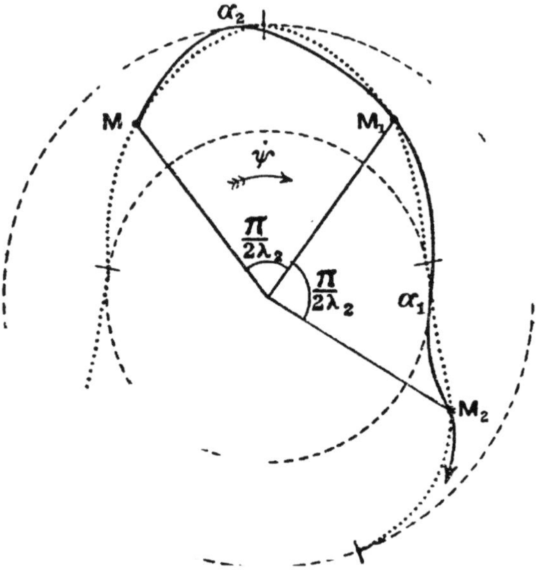

FIG. 75.

Now as θ increases beyond its mean value $\dot{\psi}$ diminishes; hence as H moves from M to a_2, the outer limit of θ, the angular distance described is less than that given by the dotted curve for corresponding values of θ. From a_2 to M_1 $\dot{\psi}$ is increasing, until at M_1 the angular distance actually described is the same as that shown by the dotted curve. From M_1 to a_1 the value of $\dot{\psi}$ is greater than the mean, and H is ahead of the dotted curve by an amount which is lost again between the positions a_1 and M_2. The whole motion is repeated after time $\frac{\pi}{\lambda_2}$.

140. Determination of the periods of normal vibrations of sleeping top. Let ξ, η, ζ, be the direction cosines of the axle referred to fixed axes OX, OY, OZ. Then since the axle is nearly vertical we have (see C. Smith, *Solid Geometry*, p. 8)

$$\xi=\sin\theta\cos\psi=\theta\cos\psi; \quad \therefore \; \dot{\xi}=\dot{\theta}\cos\psi-\theta\dot{\psi}\sin\psi.$$
$$\eta=\sin\theta\sin\psi=\theta\sin\psi; \quad \dot{\eta}=\dot{\theta}\sin\psi+\theta\dot{\psi}\cos\psi.$$
$$\zeta=1.$$

With the usual notation, let the axes (1), (2), (3) be drawn so that OZ and (3) are nearly coincident, and the angle between OX and (1) is ψ, that between OY and (1) being $\left(\frac{\pi}{2}-\psi\right)$.

The angular velocities about (1), (2), (3) are

$$-\dot{\psi}\sin\theta,\ \dot{\theta},\ \omega \text{ respectively.}$$

The angular momenta about (1), (2), (3) are

$$-A\dot{\psi}\sin\theta,\ A\dot{\theta},\ C\omega \text{ respectively,}$$

$$\text{or}\quad -A\dot{\psi}\theta,\ A\dot{\theta},\ C\omega.$$

The angular momentum about OX

$$=(-A\dot{\psi}\theta)\cos\psi+A\dot{\theta}\cos\left(\frac{\pi}{2}+\psi\right)+C\omega\,.\,\xi$$

$$=-A\dot{\eta}+C\omega\xi.$$

Similarly about OY

$$=A\dot{\xi}+C\omega\eta.$$

The torques about these axes are $-Mag\eta$, $Mag\xi$ respectively.

Hence, since the rate of change of momentum is equal to the torque, and OX, OY are fixed in space,

$$-A\ddot{\eta}+C\omega\dot{\xi}=-Mag\eta,$$

$$A\ddot{\xi}+C\omega\dot{\eta}=Mag\xi;$$

whence

$$\left.\begin{aligned}&\ddot{\xi}+2\lambda_1\dot{\eta}-(\lambda_1^2-\lambda_2^2)\xi=0,\\ \text{and}\quad&\ddot{\eta}-2\lambda^1\dot{\xi}-(\lambda_1^2-\lambda_2^2)\eta=0.\end{aligned}\right\}$$

The equations can both be satisfied by putting

$$\xi=r\cos(\lambda t+\epsilon)\ \text{and}\ \eta=r\sin(\lambda t+\epsilon),$$

provided that

$$\lambda^2-2\lambda_1\lambda+\lambda_1^2-\lambda_2^2=0,$$

$$\textit{i.e.}\quad (\lambda-\lambda_1)^2=\lambda_2^2,$$

$$\text{or}\quad \lambda=\lambda_1\pm\lambda_2,$$

which shows that the general solution of ξ and η is

$$\xi=r_1\cos\{\overline{\lambda_1+\lambda_2}\,.\,t+\epsilon_1\}+r_2\{\cos\overline{\lambda_1-\lambda_2}\,.\,t+\epsilon_2\},$$

$$\eta=r_1\sin\{\overline{\lambda_1+\lambda_2}\,.\,t+\epsilon_1\}+r_2\{\sin\overline{\lambda_1-\lambda_2}\,.\,t+\epsilon_2\}.$$

Hence the motion is the combination of two oscillations of periods $\frac{2\pi}{\lambda_1\pm\lambda_2}$.

141. Geometrical interpretation of the two oscillations. The coordinates of the projection of H on the plane XY are proportional to ξ and η.

Referring now to Fig. 56, p. 104, and substituting λ_1, λ_2 for ω_1 and ω_2 respectively, we see that P describes a circle uniformly with velocity $\lambda_1+\lambda_2$, while the image of P in the line ZR_2

describes a circle with uniform velocity $\lambda_1-\lambda_2$. Moreover ZN is $ZP\cos(\lambda_1\pm\lambda_2)t$ and PN is $ZP\sin(\lambda_1\pm\lambda_2)t$, according as we consider the point P or its image.

Hence if we take two circles of radius $2r_1$, $2r_2$, described with uniform velocities $(\lambda_1+\lambda_2)$ and $(\lambda_1-\lambda_2)$ respectively, by particles P_1, P_2, whose phases differ by $\epsilon_1-\epsilon_2$, the coordinates of H, the middle point of P_1P_2, referred to fixed axes suitably chosen, are the values found above for ξ and η respectively. Hence the locus of H gives the projection on the plane XY of the path of the head of the top in space.

142. To determine the value of ψ at any instant. Combining equations (iii) and (iv) of Art. 135 and writing $(2\lambda_2 t+2\delta)=\chi$, we have

$$\dot{\psi}=\lambda_1+\frac{\beta\lambda_1}{K+k\cos\chi};$$

$$\psi+L=\lambda_1 t+\beta\lambda_1\int\frac{dt}{K+k\cos\chi}$$

$$=\lambda_1 t+\frac{\beta\lambda_1}{2\lambda_2}\int\frac{d\chi}{K+k\cos\chi}$$

$$=\lambda_1 t+\frac{\beta\lambda_1}{2\lambda_2}\cdot\frac{2}{\sqrt{K^2-k^2}}\tan^{-1}\left\{\sqrt{\frac{K-k}{K+k}}\cos\frac{\chi}{2}\right\}$$

(see Williamson, *Integral Calculus*, p. 19);
or employing equation (vi), Art. 137, and substituting for K, k, and χ,

$$\psi+L=\lambda_1 t+\tan^{-1}\left\{\frac{a_1}{a_2}\cos(\lambda_2 t+\delta)\right\}.$$

The constant L can be easily determined from the initial conditions.

MISCELLANEOUS EXAMPLES.

1. A fly-wheel is mounted on an axle so that its plane makes an angle α with the axle: show that there is a couple on the bearings of amount $(C-A)\Omega^2\sin\alpha\cos\alpha$ where Ω is the angular velocity about the axle.

2. A conical shell of angle α is made to rotate with uniform angular velocity Ω about a generator, which is vertical. Show that a perfectly rough sphere, of radius a and radius of gyration about a diameter k, can remain in contact with the inner surface of the shell, always at the same point, in a state of steady motion, if its angular velocity about the common normal be

$$\left(1+\frac{a^2}{k^2}\right)\Omega\sin\alpha-\frac{aR}{k^2}\cdot\Omega\sin\alpha\tan\alpha+\frac{ga}{k^2\Omega},$$

where R is the distance of the point of contact from the vertex.

(Coll. Exam.)

3. A rough inclined plane is made to rotate about a normal axis with constant angular velocity, and a sphere is given any spin and placed upon it. Show that, whatever be the inclination of the plane, the angular velocity of the sphere about the normal to the plane through the point of contact remains constant.

4. A heterogeneous sphere has its centre of gravity G at a distance c from the centre O of the sphere, and the radius of gyration about any line through G is R. The sphere is placed on a fixed smooth horizontal plane, spinning with angular velocity Ω about a radius inclined at an angle α to the vertical; and at the instant of release the axis OG is in the vertical plane containing the axis of rotation and makes an angle β with that radius. Show that G describes a horizontal straight line if

$$R^2\Omega^2 \sin\alpha \sin\beta = gc \,.\, \sin^2\overline{\alpha+\beta}.$$

5. A shell in the form of a prolate spheroid whose centre of gravity is at its centre contains a symmetrical gyrostat, which rotates with angular velocity ω about its axis, and whose centre and axis coincide with those of the spheroid. Show that in the steady motion of the spheroid on a perfectly rough horizontal plane when its centre describes a circle of radius c with angular velocity Ω, the inclination α of the axis to the vertical is given by

$$\{Mbc(a\cot\alpha+b)-Ab\cos\alpha+C(a\sin\alpha+c)\}\Omega^2+C'b\omega\Omega-Mgb(a-b\cot\alpha)=0,$$

where M is the mass of the shell and the gyrostat, A the moment of inertia of the shell and the gyrostat together about a line through their centre perpendicular to their axis; C, C' those of the shell and gyrostat respectively about the axis, a the distance measured parallel to the axis of the point of contact of the shell and plane from the centre, and b its distance from the axis. (Camb. Math. Tripos.)

6. A sphere whose diameter is equal to the difference of the radii of two spherical shells is placed between them, while each shell is made to rotate with uniform angular velocity about some fixed axis through their common centre, though the angular velocity and axis need not be the same in each case. Show that the centre of the sphere will describe a circle uniformly, provided there is no slipping.

7. A rough inclined plane is driven round a vertical axis with constant angular velocity, and a sphere is placed on it. Show that, if the plane is inclined to the vertical at an angle greater than about 18°, steady motion is possible with the centre of the sphere describing a circle in space.

8. A rough horizontal disc can turn about an axis perpendicular to its plane, and a right circular cone, vertical angle α, rests on the disc with its vertex at the axis. If the disc be made to rotate with angular velocity Ω, show that the cone takes up an amount of kinetic energy equal to

$$\frac{1}{2}\Omega^2\Big/\left(\frac{\cos^2\alpha}{A}+\frac{\sin^2\alpha}{C}\right),$$

C and A being respectively the moments of inertia of the cone about its axis and a line perpendicular to the axis through the vertex.

9. A uniform solid right circular cone of vertical angle 2β is placed on a rough inclined plane of slope α, so that the generator in contact is horizontal. Prove that the cone will always be in contact with the plane, and will oscillate through two right angles, if

$$\tan\alpha < \frac{1+3\sin^2\beta}{9\sin\beta\cos\beta}.$$

10. A homogeneous right circular cone, of which the C.G. is at the distance h from the vertex and the semi-vertical angle is α, rolls on a rough inclined plane, starting from rest when it touches the plane along a horizontal line. Prove that, when the generator in contact makes an angle ψ with the horizontal lines in the plane,

$$I\dot{\psi}^2 = 2mgh\sin^2\alpha\sec\alpha\sin\gamma\sin\psi,$$

and that the point of action of the resultant normal reaction between the plane and the cone is at a distance

$$h\{\cos\alpha+3\sin\alpha\tan\gamma\sin\psi\}$$

from the vertex; where I is the moment of inertia about a generator, and γ is the inclination of the plane to the horizontal. (Coll. Exam.)

11. A solid uniform prolate spheroid whose axes are $2a$, $2b$, $2b$, spins steadily on a smooth horizontal table. It has angular velocity n about its axis of figure, that axis has angular velocity ω about the vertical, and h is the constant height of the centre above the table. Show that

$$n^2 \geqslant \frac{5(a^2+b^2)(h^2-b^2)}{b^4}\cdot\frac{g}{h},$$

and that, if n has its least value,

$$\omega^2=\frac{5(a^2-b^2)}{a^2+b^2}\cdot\frac{g}{h}.$$ (Camb. Math. Tripos)

12. A uniform solid sphere of radius c rolls under gravity in contact with a perfectly rough elliptic wire of semi-axes a and b, whose plane is horizontal: the centre of the sphere moving in a vertical plane through the major axis of the ellipse. Prove that if ω be the angular velocity of the sphere when its centre is at a height z above the major axis,

$$\omega^2=\frac{2gb^6(h-z)}{b^6(5z^2+2c^2)+5(z^2+c^2-b^2)(a^2-b^2)^3},$$

the value of z when $\omega=0$ being h, and $c>b$.

13. A circular disc has a thin rod pushed through its centre perpendicular to its plane, the length of the rod being equal to the radius of the disc. Prove that the system cannot spin with the rod vertical, unless the velocity of a point on the circumference of the disc is greater than the velocity acquired by a body after falling from rest vertically through a height ten times the radius of the disc. (Coll. Exam.)

14. A wheel with $4n$ spokes arranged symmetrically rolls with its axis horizontal on a perfectly rough horizontal plane. If the wheel and spokes be made of a very fine heavy wire, prove that the condition for stability is

$$V^2>\frac{3}{4}\,\frac{2n+\pi}{4n+3\pi}ga,$$

where a is the radius of the wheel and V its velocity. (Coll. Exam.)

15. Two light rods OP, PQ, each of length $2a$, are smoothly jointed at P, and are the axes of equal gyrostats whose centres of mass are at the middle points of the rods. The gyrostats spin with equal angular velocities n in such directions that both would spin in the same way if OPQ were a straight line. O is fixed and Q slides above O on a smooth vertical rod OZ. If M is the mass of each gyrostat, A and C its principal moments of inertia, and a mass m is suspended from Q, show that steady motion is possible with a precession Ω, in the same sense as the resolved part of any angular velocity n along OZ, provided that $k-l$ lies between unity and zero, where

$$k=\frac{Cn}{\Omega(A+Ma^2)}\quad\text{and}\quad l=\frac{2(M+m)ga}{\Omega^2(A+Ma^2)}.$$

Show further that the motion is always stable. (Camb. Math. Tripos.)

16. Prove that the least velocity v with which a thin circular disc (radius a) must be started in order to roll steadily on a rough horizontal plane in a straight line, or very nearly in a straight line, is given by $v^2>\frac{1}{3}ga$; and that the period of a small oscillation is

$$2\pi\left\{\frac{A(A+Ma^2)}{C\omega^2(C+Ma^2)-MgaA}\right\}^{\frac{1}{2}}.$$

17. A solid of revolution has an equatorial plane of symmetry, and is rolling with angular velocity ω round its axis in steady motion on a perfectly rough horizontal plane, the equatorial plane of the solid being vertical.

This motion being slightly disturbed, prove that the period of vibration is

$$2\pi\left\{\frac{A(A+Ma^2)}{C\omega^2(C+Ma^2)-Mg(a-\rho)A}\right\}^{\frac{1}{2}},$$

where ρ is the radius of curvature of the meridian of the solid at the equator, and a the radius of its equatorial circle.

18. A sphere of mass m and radius a contains a symmetrical gyrostat, mass M, freely pivoted on a diameter of the sphere. The latter is spinning on a rough horizontal plane with angular velocity Ω about this diameter, which is vertical; and Ω' is the angular velocity of the gyrostat. Show that in a small oscillation the point of contact describes an ellipse in time $\frac{2\pi}{n}$ where

$$n\{(M+m)a^2+mk^2+A\}=(\overline{M+m}a^2+mk^2)\Omega+C\Omega',$$

A and C being the moments of inertia of the gyrostat, and k the radius of gyration for the sphere about a diameter.

19. A top, free to turn about a fixed point on its axis, at which the principal moments of inertia are A, A, C and whose distance from the centre of gravity is h, is started when its axis makes angle $\frac{\pi}{3}$ with the vertical drawn upwards so that the spin about the axis is $\frac{A}{C}\left(\frac{3Mgh}{A}\right)^{\frac{1}{2}}$ and the angular velocity of its axis in azimuth is $2\left(\frac{Mgh}{3A}\right)^{\frac{1}{2}}$, the velocity in the meridian plane being zero. Show that the inclination θ of its axis to the vertical at any time is given by the equation

$$\sec\theta=1+\operatorname{sech}\left\{\left(\frac{Mgh}{A}\right)^{\frac{1}{2}}t\right\},$$

so that the axis continually approaches the vertical without ever reaching it.

(Coll. Exam.)

APPENDIX.

Explanation of the precession of a wheel from conservation of angular momentum. If we have a particle *P* constrained to revolve about a fixed axis *OC*, then, by the principle of conservation of angular momentum, the nearer it gets to this axis the faster it must revolve, and *vice versâ*. (We may imagine, for instance, the particle *P*, Fig. A, to be a heavy bead sliding on a wire *OA* which is turning about *OC*.)

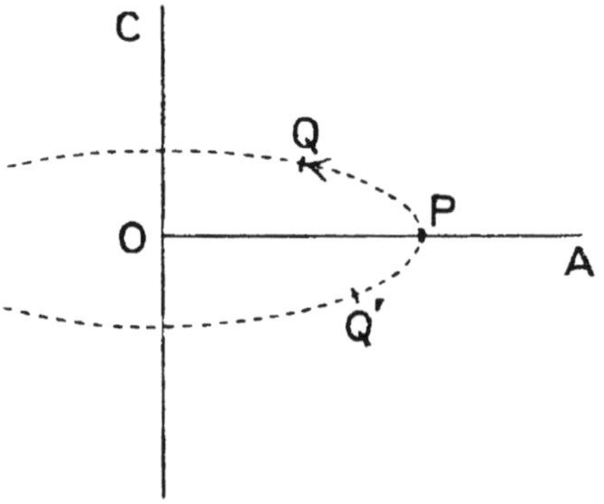

FIG. A.

This acceleration or retardation of rotation represents the action of a single force * in the appropriate direction on the particle *P*.

Thus, if the particle is getting nearer to the axis, its accelerated rotation is caused by the action of a certain force in the direction *PQ* of its rotation; while if the particle is moving away from *OC* this force acts in the direction *PQ'* opposing its rotation.

Now let us apply this.

We may consider the rotating wheel to be composed of sets of four particles, of which *A*, *B*, *C*, *D* (Fig. B) are a type, situated symmetrically as shown in the figure, and rigidly attached to one another and to the axle *OX* (perpendicular to the plane of the paper) about which they are rotating in a direction shown by the arrow-heads.

* There will really be a number of forces acting on the particle *P* due to the action of the various constraints by which it is attached to the frame. The above force is a single force equivalent to this set of forces.

On this system we shall suppose a couple whose axis is YOY' to act in such a direction that it tends to turn A and B out of the paper towards the observer, and C and D into the paper away from the observer.

We shall show that any such rotation set up by this couple results in a set of forces being called into play which tend to make the system precess about the third perpendicular axis OZ.

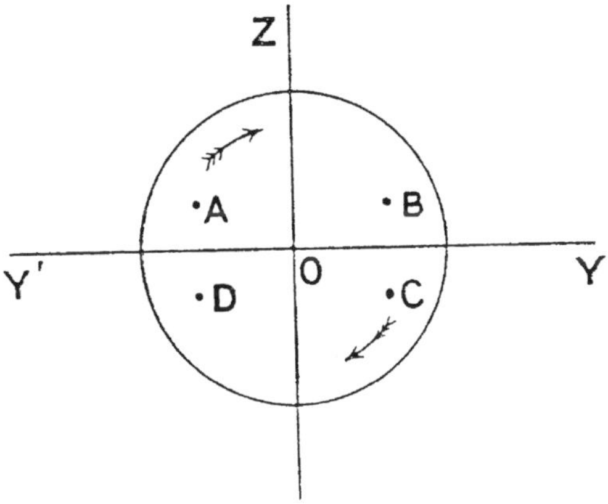

Fig. B.

By reason of its rotation about OX the particle A is increasing its distance from the axis OY'.

Thus, if it has any rotation about this axis due to the couple, there will be a tendency to diminish the speed of this rotation in consequence of the increasing distance of the particle from this axis.

This retarding tendency may be represented by a force acting on the particle A perpendicular to the plane of the paper and *away* from the observer.

Again, the particle B is getting nearer to the axis OY of the couple and so must be increasing its rate of rotation about this axis. This acceleration of rotation may be supposed to be caused by a force through B perpendicular to the plane of the paper and *towards* the observer.

Similarly, the effects on the particles C and D may be supposed caused by forces acting on C, D perpendicular to the plane of the paper and respectively *towards* and *from* the observer, since the rotation about YOY' is causing C and D to move *away* from the observer.

We see that we thus have forces at A and D away from the observer, and at B and C towards the observer.

Clearly these forces will make the system precess about the third axis OZ.

Thus, when the body is rotating about OX, the application of a couple about YO makes the body turn about the third axis OZ.

ANSWERS TO EXAMPLES.

CHAPTER I.

PAGE 16.

1. $6\frac{1}{4}$ rad./sec.^{2}. **2.** 800 radians; 2400 radians. **3.** 124950 revolutions.

PAGE 28.

1. (i) $\frac{3}{8} \cdot \frac{1}{8^2} . 10\pi = \frac{165}{896}$ lb.-ft.-sec. units.

(ii) $\frac{3}{16} \cdot \frac{1}{8^2} . 100\pi^2 = 2\frac{28803}{31136}$ lb.-ft.-sec. units.

2. (i) $\frac{7}{32} \cdot \frac{1}{24^2} \cdot \frac{200}{3}\pi = \frac{275}{3456}$ lb.-ft.-sec. units.

(ii) $\frac{7}{32} \cdot \frac{1}{24^2} \left(\frac{200\pi}{3}\right)^2 = 16\frac{30449}{45536}$ lb.-ft.-sec. units.

3. 81 : 2 : 9.

4. $\frac{10Mk^2\omega}{a}$ lb.-ft.-sec. units.

5. $3591\frac{41}{49}$ poundals.

6. Angular momentum $= \frac{7}{144}$ lb.-ft.-sec. units.
Energy $-\frac{7}{4}$ ft.-poundals.

7. $\frac{9}{\pi}$ seconds.

8. (i) 480 ft.-poundals. (ii) $128\sqrt{3}$ rad./sec.

9. 4 times.

10. (i) $\frac{8831 . \pi^2}{32 . 4 . 36} = 630\frac{5}{36}$ ft.-lbs.

(ii) $\frac{8831 . \pi^2}{32 . 36 . 15} = 611\frac{1}{9}$ lb.-wt.

(iii) $\frac{55}{144^2} = \cdot 00265$ ft.-poundals.

11. (i) $nPAl \sin\theta \cos\theta$ ft.-lbs.

(ii) $\frac{11}{9216} \frac{k^2}{nPlA \sin\theta \cos\theta}$ secs.

CORRIGENDA.

ANSWERS TO EXAMPLES.

PAGE 136.

10. (i) $\frac{8831 \, . \, \pi^2}{32 \, . \, 4 \, . \, 36} = 19$ ft.-lbs. (approx.).

(ii) $\frac{1}{2} \cdot \frac{7}{4} \cdot \frac{25}{16} \cdot \frac{6400 \, . \, \pi^2}{144 \, . \, 32} \div \frac{5}{16} \pi = 19\frac{7}{72}$ lb.-wt. $\left(\pi = \frac{22}{7}\right)$.

11. (i) $nPAl \sin\theta \cos\theta$ inch-lbs.

(ii) $\frac{11}{768} \frac{k^2}{nPAl \sin\theta \cos\theta}$ secs.

PAGE 137.

15. (i) $\frac{75 \, . \, 32 \, . \sqrt{21}}{11}$ ft.-lb.-sec. units.

(ii) 48384 ft.-poundals.

(iii) $\frac{48\sqrt{21}}{11} = 20$ secs. approx.

16. $k^2 - \frac{11}{3}$ in.^{2}.

12. Because a finite force cannot be produced instantaneously, and therefore some time must elapse before the maximum pull is attained. At the first pull, a smaller force, which is increasing up to a maximum, takes its share in producing momentum ; but later this inferior force is taken up in destroying the existing momentum in the opposite direction to it, and by the time that it is again producing momentum, the maximum force has been attained, and thus the total work done in producing momentum = the maximum pull × the length of the string. Thus the energy, and therefore the momentum produced are greater in the second case.

13. The ratio of the energies produced—*i.e.* the work done in the two cases, by pulling a string of length l with a force T, is

$$\left(T-\frac{T}{e^{2\pi\mu}}\right)l : Tl$$

$$-e^{2\pi\mu}-1 : e^{2\pi\mu}.$$

14. (i) 1000 ft.-lb.-sec. (ii) 40400 ft.-poundals.
(iii) 16 rad./sec. (iv) $12\frac{1}{2}$ seconds.

15. (i) $218\frac{2}{11}$ ft.-lb.-sec. units. (ii) 2304 ft.-poundals.
(iii) $1\frac{1}{11}$ seconds.

16. $k^2=\frac{11}{3}$ ft.^{2}

17. $\frac{550.400.121}{49}\div\frac{550.605}{7}=11\frac{3}{7}$ H.P.

$\frac{550.400.121}{49}\div\frac{2750.44}{7}=31\frac{3}{7}$ ft.-lbs.

18. (i) $96F$ ft.-poundals.

(ii) (α) $\frac{9\lambda}{2}$, λ being the constant of proportion.

(β) $\frac{9\lambda}{2}$.

The same forces are used in each case, but in inverse order, and hence the result is the same.

(iii) 9λ. (iv) $\frac{3^{n+1}}{n+1}\lambda$, $\omega=\sqrt{\frac{2(KE)}{I}}$.

CHAPTER II.

PAGE 42.

1. $156\frac{1}{4}$ ft.-lbs. **2.** $\frac{7}{2750}$ rad./sec. **3.** 10 cm.
4. 468 rad./sec. **5.** $\sqrt{2}$ feet.

PAGE 45.

1. 43·816 ft./sec. **2.** The wheels on the inside of the curve.

3. The radius $=\sqrt{\frac{2ag}{\omega\Omega}}$. **4.** $\frac{1}{2}Ma^2.\frac{\omega V}{R}$ ft.-poundals.

5. $\frac{11200.\pi}{297}=118\frac{14}{27}$ ft.-lbs. **6.** ·41 inches.

7. (*a*) By the internal stresses.

(*b*) About a vertical axis in a direction which would turn the back wheel towards the centre of the curve, and the head of the machine outwards.

(*c*) The reaction causes the back wheel to be turned away from the centre of the curve, and the head of the machine inwards, thus increasing the risk of skidding.

(*d*) $\frac{80 . \pi}{27} = 9 \cdot 3$ ft.-lbs. approximately.

CHAPTER VI.

Page 95.

5. $(A+3Ma^2)\Omega^2 \cos \alpha - C\omega\Omega = (3M+M')ag.$

(*a*) Yes; since their axes are merely translated and not rotated, it does not matter whether they are spinning or not.

(*b*) No; the figure would become skew.

6. $C_1\omega_1\Omega - (A_1 + M_1a^2 + M_2\overline{4a^2+2ab})\Omega^2 \cos \alpha = (M_1+2M_2)ag.$

$C_2\omega_2\Omega - (A_2 + M_2\overline{b^2+2ab})\Omega^2 \cos \alpha = M_2bg.$

(i) No. (ii) Yes. (iii) No.

7. The same way.

ANSWERS TO QUESTIONS.

CHAPTER I.

[1.] (α) $[M][L]^2$ (β) $[M][L]^2[T]^{-1}$. (γ) $[M][L]^2[T]^{-2}$.

[2.] 448 lb.-ft.^{2}. [3.] It is involved in the mass *M*. [4] 40π.

[5.] $Pt=M(v_1-v)$. [6.] $K.E=\frac{1}{2}Mv^2$. [7.] $Ps=\frac{1}{2}M(v_1^2-v^2)$.

CHAPTER II.

[8.] (i) It upsets it.
(ii) It causes it to precess.

[9.] (i) The further end dips.
(ii) The further end turns to the left.
(iii) The near end dips further.

[10.] (*a*) The front of the car is depressed.
(*b*) The back of the car is depressed.

[11.] (*a*) The bows turn to starboard.
(*b*) She heels to port.

[12.] The "Centrifugal Force."

CHAPTER III.

[13.] There is a small frictional couple at the toe, due to the toe not being a mathematical point.

[14.] Because if the top begins to lean over, the gravity-couple about the edge of support causes it to precess; and if the precession is hindered by a *rough* surface, the top falls down.

[15.] There is more friction to "hurry" the precession.

[16.] The radius a is greater and thus ω is reduced at a greater rate.

[17.] The final direction of friction is opposite to its initial direction.

[18.] In both cases the result of the air is the same. But if the peg is blunt the effect of friction mentioned in question 13 is more marked than in the case of the fine peg. On the other hand, friction acts at a larger arm in the case of the blunt peg, so that a smaller frictional force is capable of restoring ω to the value necessary for steady motion.

[19] It would move in a vertical straight line; or if the motion were steady it would be at rest.

[20.] Yes. The axle would describe a cone with vertex G.

[21.] $a\omega < A\Omega$. The heavier the top, the better it fulfils this condition; for we have a big torque acting, and consequently a large Ω.

[22.] The motion becomes steady more quickly. For the large spin of the body is only partially communicated to the loose spindle; hence the $a\omega$ (of the spindle) is never very large, and soon $= A\Omega$.

CHAPTER IV.

[23.] No. For since it has no angular momentum about its axis, there will be no horizontal impulse on the cushion.

CHAPTER VI.

[24.] No. For points of the body on the momentum-axis are not instantaneously at rest.

[25.] $\omega \cos \beta + \Omega \sin \alpha \sin \beta$.

[26.] (i) If convex, the direction of R will not allow G to move in a horizontal circle, so that steady motion is impossible.

(ii) If concave, steady motion will be possible. The direction of precession will depend on which side of the axle the reaction R lies, *i.e.* on the angle at which the top spins and the direction of the common normal at the point of contact.

[27.] Zero, during steady motion.

[28.] Either Ω would be zero; or the figure would fold up about OE, according to the nature of the joints at O and E.

[29.] The torque of Mg and R about D; for the vertical reaction of the plane on the cone, although equal to Mg, does not act through the line of action of the latter.

GLASGOW: PRINTED AT THE UNIVERSITY PRESS BY ROBERT MACLEHOSE AND CO. LTD.

Models of the Spinning Tops mentioned in this Chapter can be obtained from Messrs. Newton & Co., Scientific Instrument Makers, 3 Fleet Street, London, E.C.

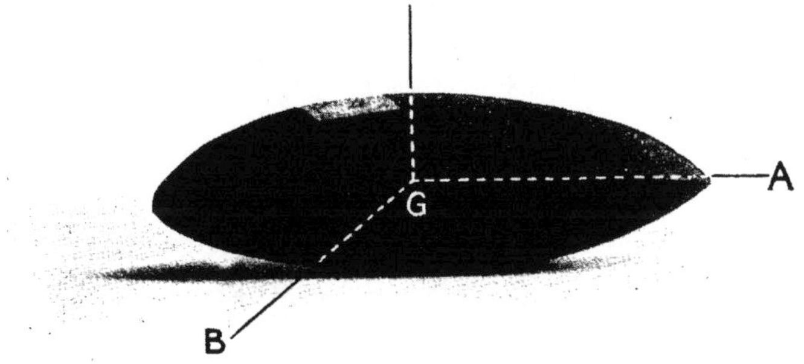

Fig. XI.

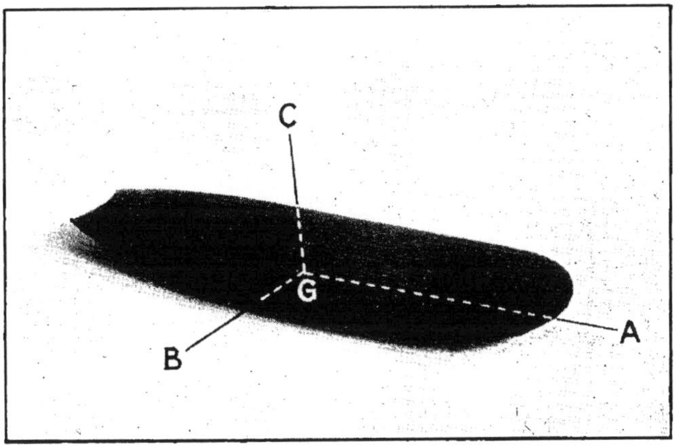

Fig. XII.

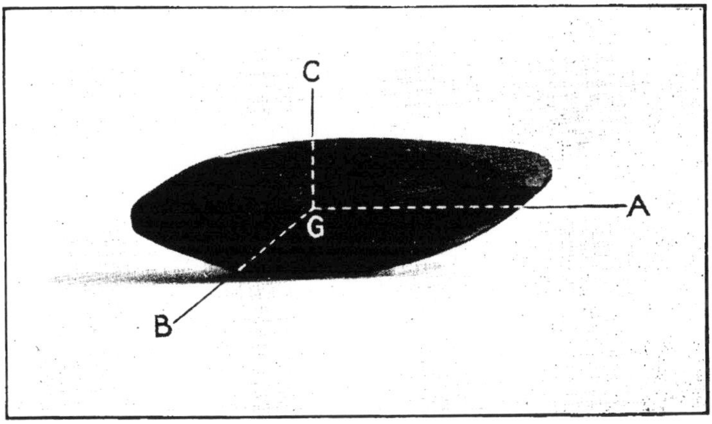

Fig. XIII.

[see pp. 7 and 54

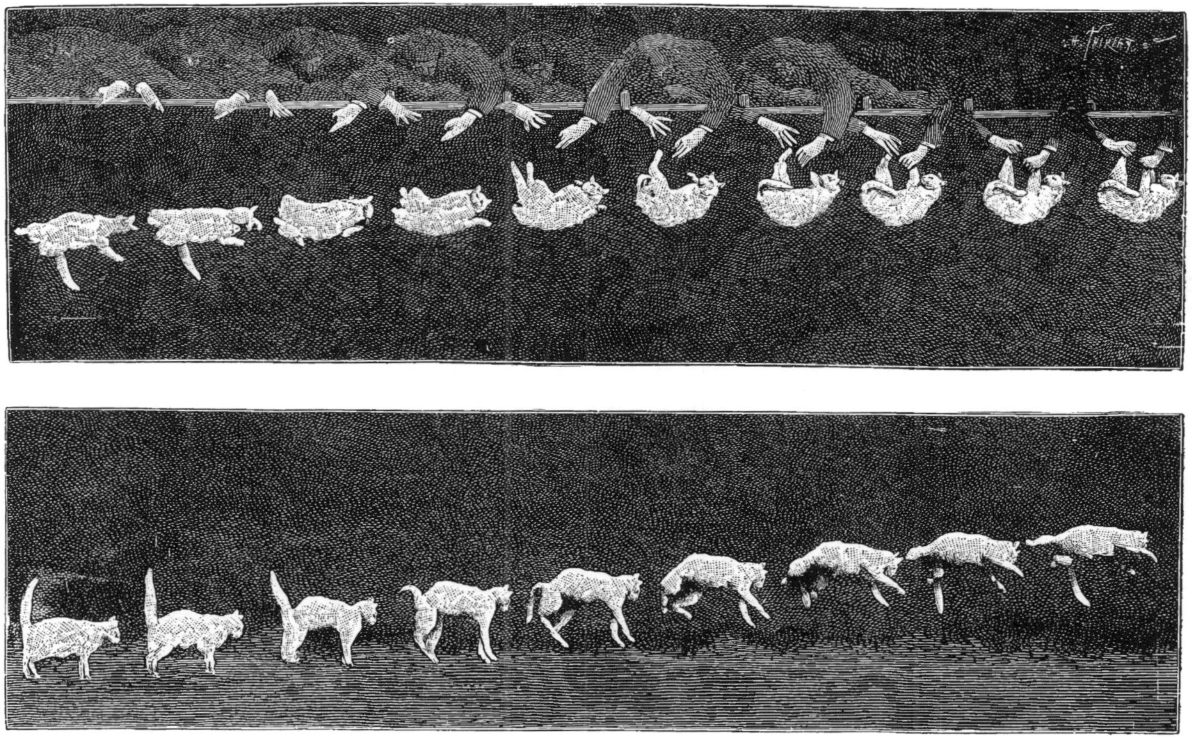

PHOTOGRAPH OF CAT RIGHTING ITSELF IN MID-AIR. (INITIAL POSITION ON THE RIGHT.)

[see p. 57.

Photo: Underwood and Underwood

SCHLICK'S APPARATUS FOR STEADYING SHIPS.

Photo: Underwood and Underwood.

MODEL EXHIBITED BEFORE THE ROYAL SOCIETY IN 1907.

[see p. 67.

Made in the USA
Middletown, DE
04 April 2020

87782246R00093